Chronicles of the Wandering Star

Written by

Kelly McCullough

Illustrated by

Carlos Lopez

Copyright 2006: It's About Time, Herff Jones Education Division

It's About Time®, and InterActions in Physical Science™ are registered trademarks of
It's About Time, Herff Jones Education Division.

Registered names and trademarks, etc., used in this publication, even without specific indication thereof,
are not to be considered unprotected by law.

All rights reserved. No part of this publication may be reproduced, stored in a retrieval system, or transmitted, in any
form or by any means, electronic, mechanical, photocopying, recording, or otherwise, without the prior written
permission of the copyright owner. Care has been taken to trace the ownership of copyright material contained
in this publication. The publisher will gladly receive any information that will rectify any reference
or credit line in subsequent editions.

Printed and bound in the United States of America.

ISBN #1-58591-405-3

1 2 3 4 5 VH 09 08 07 06

This project was supported, in part, by the National Science Foundation.
Opinions expressed are those of the authors and not necessarily those of the National Science Foundation.

© It's About Time

Table of Contents

© It's About Time

Landing on the Snack Tray

The *Wandering Star* was a Trade Confederation deep-space exploration vessel on its way home. It was preparing for Fold Space breakthrough. As the ship reached 60 percent of light speed, the Fold Space Lance began its charge cycle. At 70 percent, the cycle finished and the four Lance heads went active. At 80, the order was given to engage the Lance. With a terrible ripping noise, the Lance heads fired. From each of the four heads a stream of energy crackled toward an invisible point in space. Where they met, a hole was punched in the fabric of the universe. It only lasted for a few seconds, but by then the *Wandering Star* was traveling at 90 percent of the speed of light. The huge ship plunged through the hole and vanished from the normal universe. Behind it, the tear in the stuff of reality silently closed up.

✳ ✳ ✳

"I really, really hate that," said Modulas. He was always very tense when the *Wandering Star* would make its Fold Space crossover.

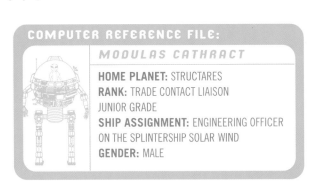

COMPUTER REFERENCE FILE:

MODULAS CATHRACT

HOME PLANET: STRUCTARES
RANK: TRADE CONTACT LIAISON JUNIOR GRADE
SHIP ASSIGNMENT: ENGINEERING OFFICER ON THE SPLINTERSHIP SOLAR WIND
GENDER: MALE

© It's About Time

"Oh, don't be a baby, Modulas," said Stas. "There have only been eight Fold Space breakthrough accidents in the 7000-year history of the Trade Confederation." After two years of exploring new star systems with the *Wandering Star*, she was just happy to be headed for home. She wasn't going to let something as unlikely as a Fold Space accident worry her.

"Besides, said Teract, "breakthrough is like the biggest rush ever invented. I love that moment just as we slide through the hole when it feels like my whole head is turning inside out." The successful breakthrough had left her feeling great.

"Actually, Teract," said Kinet, "I think that's the feeling that makes Modulas so unhappy. Being turned inside out, even for a little while, can upset your stomach." Now that the transition was over, he was impatiently hoping for something new to do.

"You don't feel that way about it, do you, Kinet?" asked Teract.

"Of course not," said Kinet. "But I'm a smallship pilot, which means I'm already a maniac."

"Good point," said Teract.

The four very different aliens formed a pod that was part of a larger Confederation research cluster assigned to the scout ship *Inquiry*.

Their main duties revolved around studying new planets and species, but everyone in the Confederation had dual roles. For breakthrough and at other times when they might be called on for space duty, the four formed the crew of the tiny splintership *Solar Wind*.

COMPUTER REFERENCE FILE:

STAS LUSTRUM

HOME PLANET: ECTOVORIA
RANK: TRADE HISTORIAN LEVEL TWO
SHIP ASSIGNMENT: RECORDS OFFICER ON THE SPLINTERSHIP SOLAR WIND
GENDER: FEMALE

COMPUTER REFERENCE FILE:

TERACT BARENDUIN

HOME PLANET: SIRCOS
RANK: TRADE CONTACT LIAISON JUNIOR GRADE
SHIP ASSIGNMENT: MISSION COMMAND OFFICER ON THE SPLINTERSHIP SOLAR WIND
GENDER: FEMALE

COMPUTER REFERENCE FILE:

KINET KINIFROUS

HOME PLANET: ANIMA
RANK: TRADE CONTACT LIAISON JUNIOR GRADE
SHIP ASSIGNMENT: PILOT ON THE SPLINTERSHIP SOLAR WIND
GENDER: MALE

COMPUTER REFERENCE FILE:

SOLAR WIND

CLASSIFICATION: SPLINTERSHIP
PURPOSE: SUPPORT AN INDEPENDENT RESEARCH UNIT FOR THE SCOUT SHIP INQUIRY

© It's About Time

✳ ✳ ✳

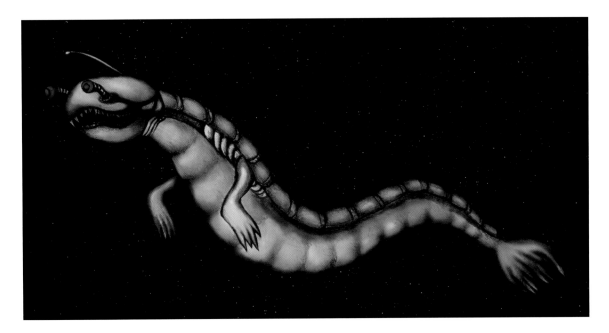

The thrask slid slowly through the depths of Fold Space. The huge beast was over 40 kilometers in length, from the front of the gaping mouth it used to mine the Foldwinds for nourishment to the tip of its massive tail. As it made its way through the depths, it was startled by the sudden tear that divided Fold Space from the normal universe. It was even more startled by the arrival of the *Wandering Star*. The Confederation ship punched through into Fold Space less than two kilometers from the mouth of the thrask. In response, the beast extended a sense organ and tasted. Slow thoughts made their way through a brain the size of a city block.

The *Wandering Star* was built almost entirely from metals and rare-earth ceramics. These kinds of materials occurred only in trace amounts in the prey animals that made up the normal diet of the thrask. The *Wandering Star* was tiny by the standards of the thrask, less than five kilometers across at the habitat ring, but it was *very* nutritious.

© It's About Time

A perfect snack. With a powerful flip of its fins, the beast thrust itself forward. The *Wandering Star* was much faster than the thrask and could turn more easily. If the ship had emerged further away from the thrask, or if its sensor officer had noticed the red light on his panel a moment sooner, nothing would have happened. But the young ensign was looking away from his board, and the ship was almost in the thrask's mouth from the start.

✳ ✳ ✳

"What in the four dimensions was that?" said Modulas, his pale, glowing skin brightening suddenly with anxiety. The *Solar Wind* was sitting in its cradle in the *Wandering Star's* hangar bay 19 with its crew aboard and ready to launch in case of emergency.

"It felt like someone hit the ship with a softening mallet," said Stas.

"*Solar Wind*, status report," said Teract, her tone firm and even.

"Splintership systems all read normal," said Kinet. "But the communication lines to *Wandering Star* control are jammed with status queries from all over. Whatever that was, it hit the whole *Wandering Star*."

The ship bucked and twisted again, bouncing in its cradle.

"It's not over yet," said Stas, clicking her mandibles sharply together for emphasis.

"I'm tapping the *Wandering Star's* engineering data stream right now," said Modulas, his fingers flying across the controls of his suit computer. "There's massive damage in Fold Lance one and two. Normal space engines are without power. Primary sensors are down. There's serious damage to virtually every system on the *Wandering Star*, but nobody seems to know what's going on."

The ship's intercom crackled to life and the captain's voice filled the cabin. "All pods prepare for crash shift. Repeat, prepare for crash shift."

"Crash shift!" clicked Stas. "That's insane. It will completely destroy what's left of the Fold Lance."

"That's better than the situation we're in now," said Modulas. "We've already lost two of the Lance heads. If we lose another one, there won't be any kind of shift. We'll be stuck in Fold Space forever."

"Modulas is right," said Kinet. "This is going to be rough. Even though the Fold Space wall is weaker from this side, two heads aren't going to make much of a hole. We may scrape some of the hull off on the sides."

"Computer, engage shock harnesses," said Teract. Padded steel arms slid out of the sides of the seats, locking the four aliens in place. "Seal suits."

Three of the four aliens pulled helmets from mounts on the sides of their seats and attached them to the collar rings of their space suits. The fourth alien, Modulas, was already wearing a full-body armored suit. Being a member of a very fragile species, he almost never left it.

"How long do you think we have before …," began Kinet.

© It's About Time

He never finished his sentence. The *Wandering Star* fired its two remaining Lance heads at that instant. The beams blasted out, striking the inside of the thrask's mouth. That wouldn't have been enough to make the beast let go. However, where the beams touched, a hole opened in the fabric of Fold Space and bits of thrask went with it. The damage wasn't enough to seriously injure the giant predator, but it did hurt. The thrask bellowed in pain, and its jaws released the *Wandering Star*. The badly wounded *Wandering Star* plunged forward through the hole in space.

© It's About Time

MODULAS CATHRACT

Rank: Trade Contact Liaison Junior Grade.

Ship assignment: Engineering officer on the splintership Solar Wind.

Gender: Male.

Modulas is from a species that is small and fragile. The flesh part of Modulas consists of a humanoid torso and a snake or fishlike lower half. He has very soft, pale skin and large eyes. He is also slightly phosphorescent.

His habitat is a small spherical unit that sits on the top of his various robotic units. The habitat has a clear dome made of extremely tough plastic, and he can be seen sitting inside. The robot unit has two arms and two legs and serves as his primary means of transportation.

Modulas is hyper-organized. The fact that his survival is contingent on the proper maintenance and operation of his suit has a lot to do with this. He is capable of

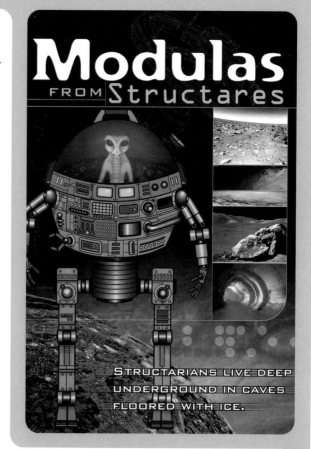

Modulas FROM Structares

STRUCTARIANS LIVE DEEP UNDERGROUND IN CAVES FLOORED WITH ICE.

spending small amounts of time outside the suit in normal Earth conditions, but he has to be very careful because of his fragile bones and low tolerance for warmth. Also, he cannot breathe the trace elements in our atmosphere for extended periods. The effects are not, however, bio-accumulative, so it is safe for him to leave his suit for brief periods. On the other hand, his atmosphere is safe for humans, so he can invite people to visit him in the larger habitat that has been set up for him and other members of his species. Structarians' suits make them very conscious of things working as integrated systems.

The suit, combined with his tendency to think everything through, tends to make him emotionally distant. However, he is deeply attached to those friends that he does make. Modulas likes to design and build things, and even enjoys maintenance tasks. He is a very logical thinker and wants things to follow logical rules. He constructs good arguments, but he doesn't like making them. One thing he does like is music. His computer is set up with a music program and he likes to compose electronic music.

His home planet is Structares, a small, cold world with relatively low gravity. It has a very bright sun and a thin ozone layer. The ultraviolet radiation on Structares is very dangerous and most organisms on the planet live underground in caves floored with ice. The ice combined with the low gravity makes for a world with very little friction. This makes getting around mostly a matter of sliding. His family is composed of his mother Menixa, stepfather Ponen, father Dulos, and brother Nodal. Like all Structarians, they are good engineers, technicians, and programmers.

© It's About Time

Look Before You Leap

The *Wandering Star* emerged into normal space in the outer reaches of a nine-planet system orbiting a small yellow star. The *Wandering Star's* radio frequency telescopes detected radio signals from the third planet out from the Sun, clear evidence of a technological society. The aliens had arrived in an inhabited system. At the time, they didn't pay much attention. They were too busy with damage control and trying to figure out what had happened to their ship.

The Fold Space gate *had* been too small. The *Wandering Star* lost her remaining Lance heads, portions of the habitat ring, most of her computer net, and all of her Fold Space communication systems. In short, she wasn't going anywhere very fast, and there was no way to phone home for help. The worst thing was that no one really knew what had happened.

After a week of emergency repairs and local space survey work, a theory about the accident was advanced. There were large amounts of frozen organic debris around the ship. In space, flesh freeze-dries in minutes. These bits of thrask suggested that they had encountered something large, alive, and hostile. The gigantic teeth marks in the skin of the *Wandering Star* tended to support that idea.

The theory explained all of the data they had, and there was a lot of evidence to support it. So the research community of the *Wandering Star* decided to accept it for the moment. Of course, if they later found new evidence that suggested a different conclusion they would revise the theory.

A couple of days later, the *Wandering Star* was ready to limp closer to the one inhabited planet, called Earth by its native population. The big question on the minds of the crew was how to contact the beings that lived there. It was decided that the *Wandering Star* should get in close, using its cloaking device to see what sort of creatures lived on the planet.

***

For almost a month, the ship hid itself in near-Earth orbit, while the computers and the researchers tried to figure out how to translate the planetary language. Unfortunately for the aliens, the people of Earth used more than one language. It was driving the crew crazy.

© It's About Time

"Hey," said Stas. "Look what I've found." Her voice was filled with excitement, and she gestured for the others to come look at her computer display. "I was just playing around with the feed from this satellite and bang, there it was."

The display showed a series of pictures along with text in the most commonly used written language, English. Strangely, probes had determined that the most commonly *spoken* language was entirely different, Chinese. It was all very confusing. Still, they had a good translator for written English. At the top of the screen was a tag that read "http://www.alphonsethefirst.com/coronation.htm."

"What is it?" asked Kinet.

"It's an account of the crowning of Alphonse the first," said Stas, "King of the Earth."

"A monarchy!" exclaimed Teract. "How primitive. This must be a very backward planet."

"Yes," said Stas, "but that doesn't matter. What matters is that this is the first evidence we've found for a planetary government. We know that there has to be one. They've got limited space travel and no one has ever found a space-faring race that didn't have a single world government."

"Good point," agreed Modulas. "If you watch the, what's the word… television, that's it. If you watch Earth television, you get the impression that these people still have countries. Now, that's a really insane notion. It's a well established fact that achieving space flight without a unified political system is completely impossible."

"Television?," said Teract. "You can't believe anything you see there. Clearly more than half of it is fiction. Stas is right, this is important. This www thing is linked to all of their research centers and these school things that they all seem to go to. It's practically their prime source of scientific information. It must be correct."

"And this data store," said Kinet, gesturing to the display, "is very clear and factual. We'll have to bump this up to Cluster Command."

"I don't know," cautioned Stas. "I mean I'm the one who found this, but where's the evidence? Shouldn't we get more supporting information first?"

"Don't be silly," replied Kinet. "This is important. We should report it as soon as possible."

✳ ✳ ✳

The discovery made by the *Solar Wind's* research pod moved quickly up the chain of command, rapidly reaching the governing council of the ship. While the ship was traveling through space, the captain had complete authority, but on matters of research and alien contact, the trade council made the choices. After a good deal of argument, they decided

© It's About Time

that this King Alphonse data supported their idea of how a planetary government should work. The council sent a delegation to Earth to make contact with its king.

Because of their part in the discovery, the crewmembers of the *Solar Wind* were included as junior members of the delegation. To any student of humanity, the results would have been predictable. Alphonse the first, whose real name was Al Smith, was completely terrified to have an 85-ton alien spaceship land in the parking lot of his apartment building. He was horror-struck that they wanted to conclude a trade pact with *him*. And he was finally reduced to hysterical laughter by the idea that anyone would believe anything they found on the web without double and triple checking it.

In the end, however, the contact team calmed him down. Al Smith suggested that they go to the United Nations and he supplied them with directions to New York and UN Headquarters. There, they were able to make contact with representatives of all of the world's governments. After a good deal of explanation and Trade Confederation diplomacy, including gifts and compliments, the *Wandering Star* concluded a mutual aid pact with the people of Earth.

❊ ❊ ❊

Meanwhile, back on the *Wandering Star*, the reputation of the crew of the Solar Wind was in tatters.

"That was just great, Stas," said Kinet raising all four arms in frustration. "You do know that we are the laughing stocks of the entire ship."

"Give her a break," said Teract. "We all agreed that the King Alphonse thing was a good idea."

"Yeah," said Modulas, "it seemed so logical at the time. All planets with space travel have planetary governments. Earth has limited space travel. Therefore, Earth must have a planetary government. Then one of us," he tapped his chest with one of his mechanized arms, "makes the big discovery. It was fantastic. Of course we all went for it."

© It's About Time

"I suppose you're right," said Kinet. "But we should have thought things through and taken a second look before we stuck our necks in the noose."

"A third and a fourth look wouldn't have hurt either," said Stas, her voice very glum. "I'm never going to do something like that again."

"Of course you will," said Kinet, some of his normal cheer returning. He was an eternal optimist. "So will I. Everybody makes mistakes, but maybe we won't make this one again."

"What I don't understand," said Modulas, "is why nobody higher up the command chain figured it out."

"Well," said Teract. "As someone who's slightly higher up that chain than you, I can suggest one reason. We all wanted it to be right. It's easy to fool yourself into believing an argument if it fits with what you want to believe. The *Wandering Star* is in rough shape, and a single planetary government would have made our job much easier. We weren't the first people to fool ourselves, and we won't be the last."

"She's right," agreed Stas. "In the computer, I can find thousands of records of people believing what they wanted to, even when the evidence didn't support it."

"Yes," said Modulas as his glow increased, a sign of strong emotion. "None of us wanted to believe that a space-faring race could still have divided government, and we forgot one of the first rules of clear thinking: 'If a theory or argument doesn't take into account all of the available information, it needs to be revised.'"

"Look," said Kinet. "I'm sure that we could all sit around discussing this for days, but I'd like to move on. Does anyone know what our Earth research topic is going to be?"

It was a sudden and unexplained shift of subject. From anyone else, it would have been jarring. From Kinet, it was normal. Teract answered without looking up from her display. "The scout ship *Inquiry* is going to start by looking at these school things. Apparently, small humans have to live through an enormous amount of information transfer to become large humans."

"That's weird," said Modulas. "Why don't they just plug them into the computer for a few weeks?"

"I didn't ask," replied Teract. "I also didn't ask about our specific assignment. The head of the *Inquiry's* research cluster made it very clear that she would prefer that I were somewhere else."

"Don't worry about it too much," said Kinet. "We just have to wait for someone else to make a big mistake and draw attention away from ours."

"That shouldn't take too long," said Stas. "These humans are strange creatures."

© It's About Time

Questions

1. At first, the aliens thought that Al Smith was King of the Earth. What information did the aliens base this conclusion on?

2. Do you think that all information on the Internet is suspect? Write your reasons.

3. Describe several things that the aliens could have done to avoid reaching a hasty conclusion about Al Smith.

4. If you had to find out some information about a set of humans, such as the Masai of Kenya, what would you do?

5. Write at least three sentences about a time when you or someone you know arrived at a hasty conclusion, and include a description of what made it a hasty conclusion. If you can't think of a real example, then make one up.

© It's About Time

COMPUTER REFERENCE FILE:

STAS LUSTRUM

Rank: Trade Historian Level Two.

Ship Assignment: Records officer on the splintership Solar Wind.

Gender: Female.

Stas is an insectoid, long and elegant, rather like a praying mantis. She has a dry, shiny, green exoskeleton. She comes from a species that values tradition and is skeptical of change. They tend to be somewhat conservative. She is really interested in the role of conservation. Stas is also very methodical and orderly. She wants to get everything right and on time. This means she sometimes has trouble thinking outside of the box.

Stas is a bit shy and takes her time making friends, but she's very considerate of the friends she does make. Stas is a vegetarian. Not all of her species are, but most eat very little meat.

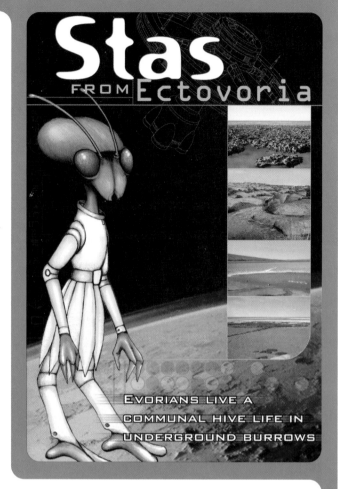

Stas
FROM Ectovoria

EVORIANS LIVE A
COMMUNAL HIVE LIFE IN
UNDERGROUND BURROWS

Her species parents in a different way. Ten sets of couples form a unit. Only one of those couples is fertile. They lay a clutch of twenty eggs, which are then divided up so that each couple raises two So, Stas has two care-parents, two birth-parents, eight aunts, eight uncles, one primary sibling (in her case a brother), and eighteen semi-siblings. She's very attached to her family. Her care-mother is Thetis, care-father Sinth, primary brother Arthis, birth-mother Sophis, birth-father Athet.

Stas comes from Ectovoria, a dry planet with no moon and no axial tilt. It's an old planet with little volcanic/tectonic activity. The Evorians are very focused on the long view. As much as possible, they tried to make their quarters resemble their stable planetary environment. Consequently, the constantly changing Earth fascinates Stas and she has many questions. Evorians are traditionally good auditors, historians, and designated worriers.

© It's About Time

What's in a Game?

The Trade Confederation Vessel *Wandering Star* has been shipwrecked in near-Earth space. The Trade Confederation is an inter-galactic organization that buys and sells information and products anywhere there is intelligent life. The *Wandering Star* is a research vessel designed to explore new planets and study their cultures. While they didn't plan on coming to Earth, or even know it existed before they arrived, the Trade Confederation researchers aren't about to miss an opportunity to study the new species called humans.

As a part of that study, the very junior crew of the tiny splintership *Solar Wind* has been assigned to visit and study middle school students for a few weeks. It's not a very glamorous assignment, but the four aliens are still young, and often have to do undesirable work. Besides, they are beginning to discover that human education is very different from what they're used to, and far more interesting.

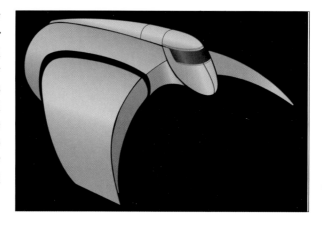

✳ ✳ ✳

"I still don't see why the humans have this whole schooling process," said Kinet. "It seems to take forever, and it's very inefficient. On Anima, we wait until the little ones are big enough to climb out of their mother's pouch, and then we give them a shot encoded with all the basic information they'll need as adolescents. Two years later, when they've reached their full growth, we give them a couple more knowledge shots and start them in their new jobs."

"The human system does seem mighty inefficient," said Modulas. "On Structares, we program everything a child needs to know into the computer of their first suit. That way they can access it any time they need it. When they outgrow the old suit, they get a new one with a better computer, and so on until they end up with an adult suit with adult programming."

"Sure it's inefficient," said Stas, "but there are cultural and biological reasons for it. Humans grow very slowly. They're one of the only species on their planet who takes more than a year or two to grow large enough to fend for themselves. It's really quite fascinating." She tilted her head to the left in thought.

© It's About Time

"I'm with Stas on this one," said Teract. "If they had a more normal educational process, our assignment would be dull as rocks. This way there's all sorts of stuff to learn. Here, take a look at this, for example."

Using her tail, she tapped a button on her computer console and an image appeared on the other three screens. It was a picture of a truly bizarre contraption that wiped the chin of a human who was eating soup. It was hideously complex, involving things as diverse as a woodpecker and an electric iron. At the top of the screen was a tag that read "http://www.rube goldberg.com/html/pencil_sharpener.htm."

"What is it?" asked Modulas. "It's an engineering nightmare."

"It's a cartoon drawn by a man named Rube Goldberg," replied Stas. "There's some background in his biography." She quickly clicked on a series of links until a page appeared with his biography. As a historian, she was fascinated by anything using the word *background*. "Let's see. Goldberg was a cartoonist, sculptor and artist in the early 1900s. His primary claim to fame came from a series of comics about Professor Lucifer Gorgonzola Butts. The professor was an inventor of unnecessarily complex devices for accomplishing simple tasks."

"I don't care what the history of it is," said Kinet. "I think it's totally stars and comets! Look at all the action happening in that contraption. I want to build one."

"Apparently, you're not alone," said Teract. "This web page talks about a contest in which Earth students build their own version of a Rube Goldberg device. There are different levels of competition ranging from middle school to college."

"I still don't get it," said Modulas. "If it takes at least 13 years of schooling to turn a small human into a full-sized one, why are they wasting any of that time on stuff like this?"

"It's all about interactions," said Teract. "Like my people, Earth's scientists use the principle of interaction to explain every kind of phenomenon that occurs."

"Sure," said Modulas. "We use interactions as well, everybody does. It's one of the basic tenets of clear thinking. 'When two objects interact, they act on or influence each other to jointly produce an effect.' It's a good system of explanation, but that doesn't account for this weirdness." He gestured emphatically at the screen.

"Think about it," said Teract. "Some interactions are obvious. In that soccer game we watched last night, the ball would get near a player. Then he would kick it. You could see his foot move into contact with the ball, hear the noise of the kick, and see the ball move in a new direction. It was clear that something was happening, but not all interactions are as obvious."

"Sure," said Stas. "If you put two magnets close together they'll move closer still and stick to each other, but you can't see why they do it."

© It's About Time

"That's a good example," agreed Teract. "There are others that are even harder to see. A puddle of water on a sunny sidewalk will slowly disappear. That's partially an interaction of the sun and the water. Or take the slow crumbling of rocks on an ocean shore. Wind and water interact with rocks to produce the effect."

"Yes," said Modulas, "evaporation and erosion. What's that got to do with this Rube Goldberg guy?"

"Everything," said Teract. "Look at all the interactions in this picture." He pointed at the cartoon. "Think of all the things you have to know about interactions to build something like this."

"Who cares what it teaches?" asked Kinet. "It would be fun to build. Let's enter."

"I'm afraid that we aren't going to be able to qualify," said Teract. "They only let humans play. Maybe if we're lucky, one of the groups of students that we're studying will sign up."

Question

Write down six examples of interactions from everyday life. Use the form:

"When _____, there is an interaction between

_____ and _____."

Example:

"When <u>a ball is kicked</u>, there is an interaction between <u>the ball</u> and <u>a person</u>."

© It's About Time

COMPUTER REFERENCE FILE:

TERACT BARENDUIN

Rank: Trade Contact Liaison Junior Grade.

Ship Assignment: Mission command officer on the splintership Solar Wind.

Gender: Female.

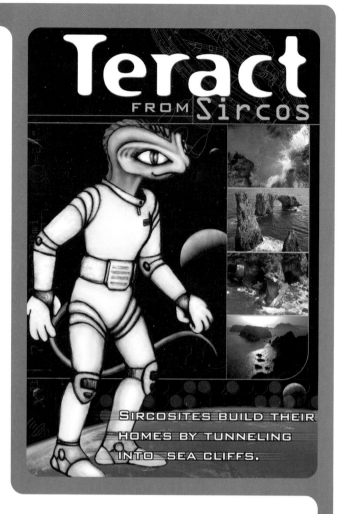

SIRCOSITES BUILD THEIR HOMES BY TUNNELING INTO SEA CLIFFS.

Teract is sleek and muscular in a compact otter-like way. She's built along swimmer/soccer player lines with a prehensile tail.

Her species is very gregarious and they traditionally live close together, so they've developed a strong interest in group interaction and social psychology. They tend to be good at soothing tempers. Teract, however, has trouble letting go of arguments, which can generate some ill will. She also has a tendency to get lost in the details, so she has a hard time looking at the big picture.

She's curious and friendly. She's also vain and very aware of her appearance. She's a middle child from a big family with three sisters and two brothers. Her family is made up of her mother, Cartha; father, Dactir; sisters, Terrat, Cret, and Rimac; brothers, Daric and Gutar.

Teract is from Sircos, a young world with active volcanos and lots of water. The Sircosites live in homes built into cliffs at the water's edge. They are agile climbers. They breathe air, but are at home both on land and in water. They can hold their breath for long periods and have internal nose flaps to facilitate swimming.

Sircosites have traditionally hunted fish with spears and nets. Their native environment with its sheer cliffs and surging tides has given them a strong cultural appreciation of the interaction between different parts of a system. Sircosites are, therefore, good mediators, diplomats, and counselors.

© It's About Time

Power Problems

The splintership *Solar Wind* was sitting nose-up on its docking cradle. When not in use, the small ship was kept in hangar 19, along with the other vessels of its research cluster. To minimize the amount of floor space it took up, the splintership was positioned on its end, with the thrusters pointed firmly at the deck.

Standing near the base of the ship were Modulas and Kinet. Since the hangar was currently pressurized, Kinet was carrying his helmet under his lower left arm instead of wearing it. As always, Modulas sat in the control section of his robotic suit. The control section, or "egg" as it was more usually called, was practically a ship's cabin in itself. It had all the same life support systems and computers. The only difference was the action unit it sat in.

Usually, that action unit was the robotlike walking suit that Modulas wore now. But the egg could be removed and placed in a number of other units, each with its own advantages and disadvantages. One of the big plusses to the walking suit was that the arms and legs it possessed were many times stronger than flesh and blood could ever be. Also, since each joint was driven by its own electrical motor, they never got tired or sore. He just had to be careful not to run out of power and to keep his systems in good shape.

This combination of strength and durability allowed Modulas to perform a lot of routine maintenance tasks on the splintership without having to resort to heavy equipment. The downside was that he wasn't very good with jobs that required a fine sense of touch. Fortunately, Kinet, with his four agile hands, more than made up for that difficulty. Together, they made a good team.

"So what's next on our list?" asked Modulas.

"Landing gear," said Kinet. "There's been a very high-pitched buzzing sound the last couple of times we deployed the landing legs. You might not have noticed it through your exterior microphones. I can barely register it, and I've got these." He waggled his long mobile ears. Like other Animites, he could hear much higher frequencies than any other known intelligent species.

© It's About Time

"That's going to be a real hassle," said Modulas. "The rear legs are easy enough to get at, but we'll have to lower the ship if you want to really go over the nose gear. And that means getting permission from hangar control."

"I know," said Kinet, "but that buzz has me worried."

"Well," said Modulas, "if you insist. Let's start with the rear systems and hope that's where the problem is. Why don't I climb up to the control cabin and run out the gear while you have a listen down here?"

"Suits me," agreed Kinet.

A few minutes later, Modulas heard Kinet's voice crackling over the cabin speakers.

"Nope," said Kinet. "I don't hear a thing."

"Too bad," sighed Modulas, "but I can't say I'm surprised. It's practically a rule of engineering that the part that's hardest to replace is always the one that needs fixing. Why don't you plug the diagnostics computer into those legs and run a full check while I argue with hangar control? We might as well fully check the rear systems while they're open."

"Can do," said Kinet.

The complete set of tests took an hour to run, and they showed that the back landing legs were fully operational. By that time, Modulas had finally gotten hangar control's permission to move the splintership onto the launch floor and to set it on its feet. The two aliens had also switched positions. Since Kinet was the pilot, he sat at the controls for the movement phase, something he had done hundreds of times before. Modulas, as the engineering officer, rode on the outside.

When the *Solar Wind* reached the open area of the hangar, in front of the exterior doors, it was time to go to work. Normally, when a splintership arrived in the launch area, a computer-controlled electro-magnetic crane lowered it onto its landing gear. Since the two aliens wanted to be able to run the front landing leg in and out to check for problems, they had to do something slightly different. Safety systems on the landing gear wouldn't allow the leg to be retracted when the ship's weight was resting on it. Instead, the ship had to be lowered almost to launch position and then held in place while a hull-metal stand was placed under the nose. It worked something like a car's jackstand.

Using his egg as a radio control rig, Modulas slowly swung the crane out over the *Solar Wind*, lowering the powerful electro-magnet until it was in position to grab hold. With the flick of a switch, Modulas powered up the magnet. There was a loud clanging noise as the surface of the magnet locked onto the hull of the *Solar Wind*. Then, just like a magnetic crane in a human junkyard lifting a wrecked car, the powerful machine raised the splintership out of its landing cradle.

© It's About Time

For a moment, the *Solar Wind* hung in mid-air, its landing legs all fully extended, before Modulas began carefully lowering it to the deck. When it was a few meters above the ground, he angled the ship so that its back legs touched down first. He put the device on hold while he wheeled the huge jackstand under the nose of the ship. This was a risky maneuver that normally would have been done by a robot. But the *Wandering Star* was still a long way from being fully repaired, and hangar control hadn't been willing to pull a maintenance robot from duty elsewhere.

Modulas had the stand almost in place when the hangar lights suddenly flickered and blinked off for a moment. With all of the work being performed on the *Wandering Star*, there was a heavy demand for electricity. More power was needed than the ship's damaged generating plant could currently supply. The flicker had been caused by a split second power loss. At any other time, that might not have been a problem. Unfortunately, the multi-ton bulk of the *Solar Wind* was suspended by a magnetic crane. An electro-magnetic crane. The power to the magnet was only off for a fraction of a second, but that was enough. The nose of the ship dropped, and Modulas was underneath.

"Ohhh graaaaaax!" Modulas screamed into his mike. Then a tremendous thud sounded in the background and he was abruptly cut off.

Modulas' voice crackling over the com had sounded terribly loud in Kinet's ears and the silence following the exclamation seemed even louder. The drop had startled Kinet but he was not hurt. He feared what it might have done to Modulas though. Kinet ripped off the safety harness holding him in the pilot's chair and dashed for the airlock at the back of the ship.

Since the hangar was currently filled with air, Kinet hit the emergency override that opened both of the airlock's doors simultaneously. The harsh sound of an alarm buzzer filled the air. Kinet bypassed the slowly extending ramp and dropped straight to the ground. Landing on all sixes, he took off for the front of the ship like a rocket. An Animite using his four arms as extra legs can really move.

At first, he couldn't see Modulas at all. The huge jackstand lay on its side next to the landing gear, blocking his view forward. Then he noticed a twisted chunk of metal extending from underneath the pad of the forward landing leg. It was the foot of Modulas' walking suit. Kinet ran so fast that he had to catch hold of the corner of the jack with two hands in order to stop, like a rollerblader grabbing onto a traffic sign.

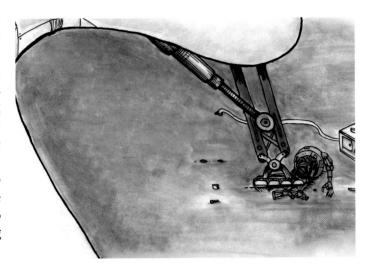

On the other side of the jack, Modulas' suit lay on its side. The first thing Kinet saw was the right knee of the suit, sticking out from under the splintership's landing leg. It was flatter than a used toothpaste tube.

© It's About Time

The tiny motors that drove the joint had been reduced to a tangle of copper wire and magnet fragments. From there, Kinet's eyes slowly made their way up the torso of the suit. He was almost too afraid to look at the egg, but that was the critical par.; As long as the life support systems were intact, there was a good chance that Modulas hadn't been killed.

Kinet looked up, and breathed a sigh of relief. The heavy armored glass of the egg was unbroken. In fact, Modulas was standing on the tilted inner surface of the egg trying to climb back into his seat. It looked like he was going to be all right. A moment later, the small glowing alien managed to pull himself up into the control chair, and his voice could be heard coming from the suit's speakers.

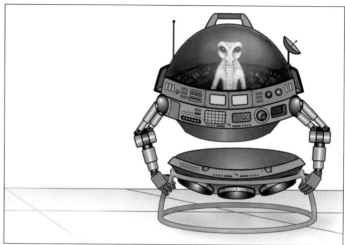

"Grax. Grax. Grax," Modulas exclaimed. He seemed to notice Kinet's presence for the first time, and his glow strengthened into a strong blush. "Oops. Sorry about that. I didn't know anyone was listening."

"That's all right," said Kinet. "I don't speak Structarian. What was that you were saying?"

"Uhh…" stammered Modulas, "drat. It means drat."

"Ohhh," said Kinet, winking. "Of course it does. Are you okay?"

"I think so," said Modulas. "I'm feeling a little bruised, and my suit is going to need a new leg at the very least, but aside from that I'm not too bad. In fact, when you consider that a spaceship just fell on me, I'm great!"

"What do you want me to do?" asked Kinet. "Should I get the ship off of you first? Or what?"

"Actually," replied Modulas, "I think it would be safer to leave it where it is until I'm out from underneath. If you could get my hovercraft action unit out here, I'll transfer my egg and then we can see about freeing the walking suit."

"Let's do that then," said Kinet.

So that he wouldn't have to leave his friend, Kinet used his com to call Teract and have her get the hover unit. Soon afterwards, Teract and Stas arrived with the suit. Using the powerful arms of the action unit, Modulas lifted himself into the new suit. Then, even though he insisted he was fine, Stas took him down to the infirmary for a thorough check up. Kinet and Teract would have gone too, but Modulas insisted that they finish the job on the landing gear.

© It's About Time

"You can't quit now," said Modulas. "The cost is too high already." He pointed at the wreckage of his walking suit for emphasis. "It took me an hour just to get permission to haul the ship out here. I don't want that time and effort going to waste. Besides, like I keep telling you, I'm perfectly fine."

There were a couple of patches on Modulas' skin where the glow had changed to a sort of sickly green. His friends thought that maybe he wasn't as well as he claimed, but they agreed to humor him. If anything was seriously wrong, Stas could call them from the infirmary.

Once Modulas was safely on his way, Kinet reattached the crane and lifted the ship. A maintenance robot, which hangar control had suddenly decided *was* a priority, placed the jackstand. After shutting off the airlock alarm, Teract ran the landing leg in and out while Kinet listened to the machinery. The buzz was still there. It turned out that a nut had worked its way partially loose from a bolt, so that when the landing gear was operated, the bolt rocked rapidly back and forth in its sleeve. The vibration from this had caused the noise. Kinet told Modulas all about it when he visited him at the infirmary that evening.

The medtechs had agreed that the little Structarian was basically unhurt, but they insisted on keeping him overnight for observation anyway. In the morning, he would be allowed to return to duty.

Questions

1. Describe four examples of electric-circuit interactions in *Power Problems*.

2. Write a five-paragraph story that includes at least two electric-circuit interactions and at least two magnetic interactions. Your story can be about the aliens or about a different topic.

© It's About Time

COMPUTER REFERENCE FILE:

KINET KINIFROUS

Kinet
FROM **Anima**

ANIMITES LIVE HIGH IN THE
TREES TO AVOID LARGE
PREDATORS BELOW...

Rank: Trade Contact Liaison Junior Grade.

Ship Assignment: Pilot on the splintership Solar Wind.

Gender: Male.

Kinet is small and a little bit on the roly-poly side. He has four arms and longish fur that is patterned in black and white. He has big eyes, a mobile rubbery nose, and long floppy ears. There's something about his appearance that suggests a combination of elements from a panda, a raccoon, and a Persian cat.

Kinet has a dynamic personality. He always wants to be in motion. In another species he might be considered hyperactive, but he comes from a highly energetic people. Especially when young, they tend to fragment and lose focus. Kinet is no exception. He's overly status conscious and really wants to be liked, so he can sometimes be talked into doing things he shouldn't. But again, he tries to be careful, because he knows that these tendencies can get him into trouble.

He makes friends easily, because he's fun to be around. Something is always happening wherever he is. He's very interested in music. Talking is Kinet's second passion after action. He speaks a mile a minute.

Kinet is an only child. One Animite child at a time is enough for most parents, so they try to space them out quite a bit, and Kinet's parents haven't yet decided whether they want another. His father is Kinear and his mother Etinara.

Kinet is from planet Anima. Anima is an older, earthlike world, with lots of water and rich soil. It's very green, with huge forests covering most of the land masses. The Animites live in the trees, which can reach heights of several hundred meters. Having four arms gives them an enormous advantage in arboreal living and Kinet can climb anything. The ground underneath tends to be swampy and is the home to many large predators. There are other predators in the trees. Dodging predators is a good part of the reason that Kinet's people are so energetic.

© It's About Time

Is The Buzzing In My Ears Bothering You?

"It'll be good to get back to my regular suit," said Modulas. "I'm getting tired of running around in this thing." He tapped the metallic hand of his suit on the steel chassis of his hover unit. He'd been forced to use it ever since a power outage had dropped a spaceship on the leg of his regular suit.

"I don't know," replied Kinet. "The hover unit is pretty cool. It's a bit like the float cars I used to race back on Anima."

"I bet your float cars were easier to steer," said Modulas. "This hover unit is a real pain. It has something like 20 different fan controls. There are power controls for the lift fan motors, and for both of the thrust fans. There's even a separate set of controls for the motors that move the fans so that I can steer."

"I thought your suit had a joystick master control," said Kinet.

"It does," said Modulas. "but it's designed for use with all of the action units. For some reason, the software just doesn't do as good a job with the hover unit. I think there may be some problem with the fan-motor interface."

The lounge door chimed. It was a tech from maintenance with Modulas' walking suit. The action unit was on a small, motorized cart. The tech rolled it in, took Modulas' electronic signature and departed with the cart. A few minutes later, Modulas made the transfer from one unit to the other.

"I'm going to try it out," said Modulas. "Want to go for a walk?"

"Sure," Kinet agreed.

As the two started to get onto the elevator, which was just a few feet from the door to their research pod's lounge, Modulas bumped into the left side of the doorway. The Structarian had to make a minor course correction to get through. Unfortunately, this took longer than a person would normally use entering an elevator and he almost got his left leg caught in the motorized door.

"Are you okay?" asked Kinet.

"Yeah," said Modulas. "I'm fine. It's just that I've been out of the suit for almost two weeks, and I'm a little clumsy."

Kinet wasn't sure he believed that. After all, Modulas had been using one walking suit or another for most of his life. It seemed unlikely that he could forget how to steer in just two weeks. He decided not to argue with his friend though.

Unfortunately, Kinet was right. When they got off the elevator, the pair set out to walk across the open park area under the main dome of the *Wandering Star*. As they walked, Modulas kept turning to the left.

"What's wrong?" asked Kinet. "Why are you walking in circles?"

© It's About Time

Modulas sighed. "I think the motors on the right side are getting more power than the ones on the left. I could probably compensate for it by steering a little right, but that would just be working around the problem. I think I'm going to have to send the unit back to the shop."

"I'm sorry," said Kinet. "I know how much you were looking forward to getting back to normal."

"Sometimes these things happen," replied Modulas. "I guess I should just be glad that it was my suit and not my legs that got mashed. The suit they can fix." He smiled sadly. "If only it didn't take so long for the shop to get things done."

Still, Kinet couldn't help noticing how down Modulas seemed. He was very quiet. The only sound on the ride back down to their pod's area was the hum of the lift motors in the elevator. Kinet resolved to do something to cheer up his friend.

"Is there a chance we could fix it?" asked Kinet.

"Maybe," replied Modulas, "but why bother. The maintenance people work on a lot of suits. I'm sure they'd do it faster."

"Sure," said Kinet, "but how long would it just sit there waiting its turn? Eighty percent of the time something is in the shop, it's just waiting for a tech to take a look at it. Even if it took us three times as long to make the actual repair, it'd probably still get done faster."

"You'd be willing to help me do that?" asked Modulas.

"Of course," Kinet replied.

"Let's try it then," said Modulas. He paused for a moment and the glow of his skin brightened briefly, indicating strong emotion. "Thanks. You're a good friend."

"Don't thank me yet," said Kinet, with a chuckle. "I might just make things worse."

✳ ✳ ✳

It had taken more than two days, but Kinet and Modulas had finally gotten the suit repaired properly. In celebration, they decided to go down to the *Solar Wind* and play some video games. The computer system on the ship wasn't any better than the one in the pod lounge, but the splintership's command chairs were wired with the best display and control hardware available. They had to be, so that the crew could keep themselves in training even when they didn't have the opportunity for live flight.

The pair had been playing a capture-the-flag game against a team on another ship for about an hour when Modulas started hearing an angry buzzing sound over his in-suit speaker system. As soon as they finished the game, he called a halt.

"Hey, Kinet, what's that noise?" asked Modulas.

"What noise?" replied Kinet. He was surprised. A sound that Modulas could hear that he couldn't would be very unusual.

"I've got a buzzing in my ears," said Modulas. "It sounds a bit like an Earth bee would sound if you put it in a jar. You can't miss it."

© It's About Time

"I'm not hearing anything," said Kinet. "Are you sure it's not part of the game?"

"Absolutely," said Modulas. "I shut down my connection to the game system, but the noise is still there."

"I hope you're not having any more suit trouble," said Kinet.

"Me too," agreed Modulas. "Especially since this would have to be in the life support egg. The sound system is completely unrelated to the action units."

"Well," said Kinet, after thinking for a moment, "let me come back to your engineering station and see if I can hear it there. Maybe it's just a really quiet sound that I can't hear from up here in the pilot's seat."

"It's worth a try," said Modulas, "but it sounds awfully loud from here."

So Kinet released his seat belts, and climbed down to engineering at the rear of the cabin. Since they were in dock, the ship was sitting on its tail in the docking cradle. But that was all right, a ladder was set into the floor and ran all the way from the pilot's station at the nose of ship to the airlock at the tail. All of the aliens were used to climbing up and down that ladder to get to their stations. It was a regular part of their duties.

Modulas was lying on his back in the engineering chair with his control systems above him. Kinet stepped from the ladder to the back wall of the cabin, next to Modulas. With the ship on its tail, the wall was actually a floor.

"Can you hear it now?" asked Modulas.

"Be quiet for a second and just let me listen," said Kinet, shushing Modulas with a couple of his hands.

Kinet slowly swiveled his big ears back and forth, trying to catch any hint of sound. After a moment, he did hear something, but it was very faint. It seemed to be coming from the armored glass dome of Modulas' egg. Leaning in very close, Kinet placed his right ear flat against the dome. He could suddenly hear the noise much better, like when you hear someone talking in the next room, but can't tell what's being said until you put an ear against the wall.

"It's coming from inside your suit," said Kinet.

"I was afraid of that," sighed Modulas. "I'd better pull the speakers and look at them."

"Shall I look at the microphones out here in case the problem is on this end?" asked Kinet.

"Good idea," agreed Modulas.

Climbing down from his seat, Modulas joined Kinet on the back wall. Sitting on the wall/floor, the pair went to work. Modulas had more familiarity with the suit's systems, so he was able to pull the first of the two internal speakers before Kinet had gotten either of the external mikes loose.

© It's About Time

Kinet paused in his own task briefly to look over Modulas' shoulder. The Structarian had folded a small working table out of the arm of his control seat in the egg and had laid the speaker on it. There wasn't really much to the speaker; just a thin plastic cone and a circular electro-magnet which could be rapidly activated and deactivated by turning the power on and off. This would set up vibrations in the cone which would come out as sound. Since there wasn't anything obviously wrong with the speaker, Kinet went back to work on the microphone.

The microphone, though much smaller than the speaker, worked on the same principle, only in reverse. Sounds hitting its tiny cone would make it vibrate. That would move the electro-magnet at its center, sending sound information in the other direction. Again, there was nothing obviously wrong, so Kinet ran some tests. After about 20 minutes, they tracked the problem to a loose connection in the left internal speaker. The bad connection had been interfering with the flow of electricity to the speaker, essentially adding a sound that really wasn't there.

They had just finished putting everything back together when Modulas heard a new buzzing sound, much louder this time. He let out a long, frustrated sigh and noticed that Kinet was covering his ears.

"Don't tell me the buzzing in my ears is bothering you now," said Modulas with a laugh.

"No," said Kinet. "It's coming from the pilot's seat. Hang on a second."

Kinet quickly climbed back up to his station. The com alert buzzer was going full volume, and a red light was flashing on the panel. Without bothering to scramble into his seat, Kinet snatched the headset that hung next to the ship's steering controls.

"Kinet here. What's the problem?"

A voice came out of the headset in response, "Are you all right?" It was Teract. "Where have you been for the last hour? I've been calling again and again, but there was no response. I was getting really worried. That's why I activated the emergency alert."

"Sorry," said Kinet. "Modulas had another suit breakdown and we've been working on that."

"Well, next time, tell someone," growled Teract. "Anyway, get your lazy carcasses up to the living quarters. I'll meet you there. I've got news."

"On our way," said Kinet.

Questions

1. Modulas's suit has at least two electromagnets in the sound system. What is an electromagnet?

2. How could the electromagnets be involved in making sound in Modulas's suit?

© It's About Time

Power Hungry

Kinet and Modulas were heading back to the pod living area after spending several hours on board the *Solar Wind*. They had been gaming and they had worked on Modulas' suit, but now, Teract needed them back to tell them something.

"You know," said Kinet, "I'd never really thought about it before, but that suit uses an awful lot of power. I guess the only reason I noticed now is all the time we've put in working on it."

"Tell me about it," agreed Modulas. "That's why the trade federation was so happy to find my people. We had the best battery technology they'd ever seen. We had to. Without lightweight, long-life, high-power batteries, we wouldn't be able to live anywhere but in the depths of the cave system on Structares."

"Speaking of which," said Kinet, pausing for a moment to open the door to the pod living quarters. "Why didn't your people have trouble generating so much power back on Structares?" The two stepped into the lounge. Stas was there, but Teract hadn't arrived yet.

"Oh, they did," said Modulas. Kinet looked confused, so he continued. "Look, a generator is just a motor in reverse, right?"

Stas looked up from her book, and joined the conversation. "Sure. Instead of transforming electrical energy into motion energy, motion energy is transformed into electrical energy."

"Exactly," said Modulas. "And how do they normally get that motion?"

"Most often with heat," replied Kinet, "by burning coal, natural gas or oil, or using a nuclear reaction they produce heat energy to make steam. The pressure of that steam turns a turbine, transferring the energy. Then the mechanical energy of the turbine is transferred again when the spinning turbine is used to produce electricity."

"Don't forget solar heat as a source for steam," said Stas. "On Ectovoria, we mostly use our sun's heat to boil water for steam."

© It's About Time

"That wouldn't work for us," said Modulas. "Structares is a small cold world, and it's a long way from our sun. We can't use fossil fuels either. Since we live deep in the caves, the exhaust from a fossil fuel plant has no place to go and that would be deadly. And, at least at first, we didn't have nuclear power."

"So what did you do?" asked Stas, clicking her mandibles with interest. She was a history buff and deeply interested in that sort of thing, but she'd never studied power generation on Structares.

"Yeah," added Kinet. "You don't really have much in the way of oceans for tidal energy like Teract's people."

"Right," said Modulas. "Since it's an old world, we don't have much geothermal heat, and most of our rivers are frozen, so we can't use them to turn turbines. Fortunately, there are a few that aren't, and that's where our society got its power when we were first developing technology, but it wasn't enough."

"How did you overcome that?" asked Stas.

"Wind," said Modulas. "Lots and lots of wind. There's practically no surface life on Structares. That's partly because of the cold, and partly because it would be blown away. So now there are wind farms all over the planet."

"But wasn't that difficult to set up?" asked Stas.

"Oh yes," said Modulas, nodding. "In addition to the cold and the wind, a lot of ultraviolet light comes through our atmosphere, and with our thin skins, that can kill us. Building the wind farms is really why we developed the suits in the first place. We needed them to be able to work on the surface. Even without the added demand of the suits, our world needed more electricity than it was making, and the only way to produce it was to build the wind farms. Things were touch and go for quite a while when they were first being built. We were just barely able to generate enough power to run the suits that we needed to build the systems needed to generate more power and so on. Thousands of my people died working on those first projects. They're considered heroes of the race, and there's a memorial to them in front of the planetary capitol."

"What does it look like?" asked Stas.

"A windmill," replied Modulas. "Of course."

© It's About Time

When Push Comes to Shove

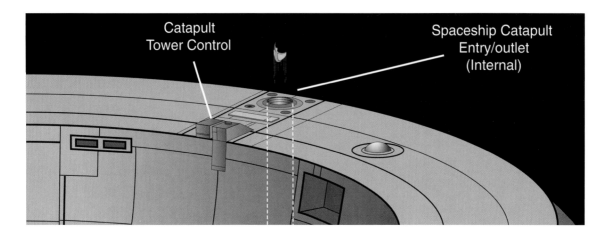

Catapult Tower Control

Spaceship Catapult Entry/outlet (Internal)

Kinet, Stas, and Modulas were waiting for Teract in the lounge area of the suite shared by their research pod. She was supposed to be bringing them some news. It was an oval room with seven doors. Four doors led to the aliens' bedrooms. One led into the small restroom they shared. One was for the drop shaft to their splintership, the *Solar Wind*. The last door opened into the larger recreation area that the pod shared with the rest of their research cluster. It was this last door that now burst open as Teract came tearing into the room.

"Woo hoo!" she shouted. "We get to take a break from research duty. We've got new orders." Her tail flicked from side to side in obvious excitement.

"What are they?" asked Modulas, his pale, glowing skin brightening to reflect the emotional intensity of his excitement.

"We're going out to the asteroids for a couple of weeks!"

"All right!" Kinet hooted, rubbing all four hands together in glee. "I'll get to do some real flying! What's up?"

"One of the reclamation and repair pods suffered a major splintership malfunction," said Teract. "They were supposed to collect orbital object 11743, but with the ship down, they won't be able to. So we got the job. Object 11743 is an asteroid rich in rare Earth metals we'll need to repair the fold space lance and our other nanocircuitry."

"How did we get picked for a job like that?" asked Stas. It was completely out of their area of specialization and the choice was a surprise.

"I'm not really sure," said Teract. "I got the impression from the folks at the Bureau of Personnel that things are pretty tight in the splintership department. Apparently, we're the only crew ready to go in the available launch window. Whatever the reason, it'll make for a nice change."

© It's About Time

"The work we're doing with the small humans is really important," said Stas. "I think it holds the key to understanding their whole culture."

Kinet nodded. "I agree completely, "but they'll still be there to study when we get back, and some off-planet duty will look good in our records for this trip."

"I suppose, but I'm still going to miss the small humans. When do we leave?"

"Yesterday at the latest. We're wanted on the catapult deck in fifty minutes."

"We'd better get moving then," said Kinet, rubbing his four hands together in anticipation. "When the Bureau of Personnel gives directions, they want them followed and you'd better stay on time and on task."

A little while later, they were all in position aboard the *Solar Wind*. The splintership's speakers crackled on.

"*Solar Wind*, this is launch control. Prepare for maximum acceleration exit in tee-minus two minutes and counting. One fifty-nine. One fifty-eight."

The aliens rechecked their acceleration straps as the voice droned on. In order to speed up their trip to the asteroid belt, the splintership was using one of the *Wandering Star's* launch catapults. Like those aboard human aircraft carriers, the catapult would push the *Solar Wind* out of her mother ship at a much greater speed than she could achieve on her own. It was a bit like shooting a paperclip with a rubber band.

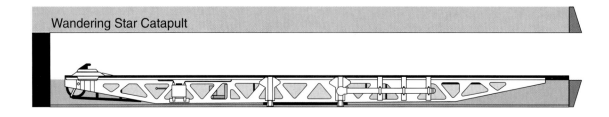

Wandering Star Catapult

With a tremendous whoosh, the splintership whipped down the launch track at a steadily increasing pace. As soon as they cleared the *Wandering Star's* safety area, Kinet turned the engines to maximum thrust. Behind them, the exhaust was an arm of fire pushing them across the night sky on course to intercept orbital object 11743.

✷✷✷

The *Solar Wind* slowly approached the asteroid. It was a delicate operation, made more difficult by the bulky load of equipment they carried for use on the asteroid. There was no significant gravity to contend with, but the ship's center of mass was not where it usually was.

© It's About Time

"Easy does it, Kinet," said Modulas. "According to my readouts, the self-propulsion kit they sent along to attach to the asteroid is off balance. I think it shifted when the launcher kicked us out."

"No worries, Mod," Kinet replied. "I could fly this boat with one hand tied behind my back."

"I'd find that more reassuring if it wouldn't still leave you with one more hand than the rest of us."

"I've got it under control." Kinet tapped the forward thrusters to bring their nose up a little more and fired the landing sequence.

The loading crew had been rushed when they attached the package to the splintership. If the *Solar Wind* missed its launch time, they would have to wait several days to be in a good position for launch again, and that would set the whole repair schedule off by a week. Because of the rush, they failed to secure all of the clamps that held the equipment in place. Already loosened by the action of the launching catapult, the remaining clamps gave way under this final stress.

Suddenly, the push that Kinet was applying to the ship with his thrusters went out of balance. The ship lurched and tumbled. Instead of the quiet thump of landing gear making gentle contact with the asteroid, there was a nasty crunch. The ship landed hard and at a bad angle.

There was no immediate rush of escaping air, so the four breathed a collective sigh of relief. They suited up and ventured out to inspect the damage.

"I don't like the looks of that at all," said Stas, pointing a hand-light at the badly bent right rear landing leg. "If we try to use that under any kind of real gravity, it's going to snap right off."

"Stas is right," said Modulas. "We won't be able to land in our normal bay, or on Earth. That's for sure."

"How are we going to deal with it then?" asked Stas.

"We'll have to worry about that later," answered Teract. "For now, let's just be thankful that they pad the mc-squared out of these engineering bundles. It looks like the self-propulsion unit for the asteroid came through all right. Let's pull it out of its box so we can give this hunk of rock a big old shove toward the *Wandering Star*."

✳ ✳ ✳

© It's About Time

It had taken the team a full three days to prepare the asteroid for its voyage into the inner system. All four aliens were happy to see it slowly heading inward guided along by the thruster pack they had attached to the asteroid. The pack would keep the asteroid speeding up until it got to the halfway point. Then it would be turned around in order to start pushing in the other direction. That way, the asteroid would slow to a stop near the *Wandering Star*.

ASTEROID TRAJECTORY PATH

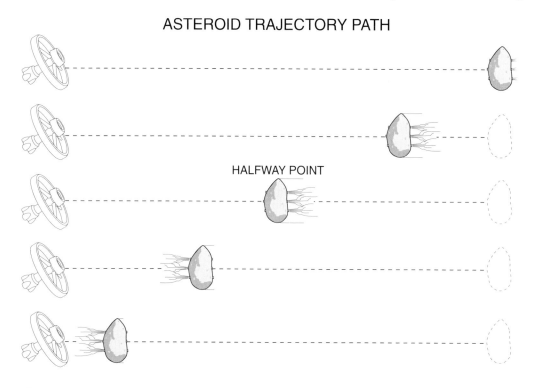

HALFWAY POINT

The aliens were not thrilled about their orders from *Wandering Star* command. The malfunctioning repair splintership was now back on-line and the Bureau of Personnel wanted them to return to their research post as quickly as possible. That meant running the rockets at a one-G acceleration all the way home and then using the launch catapult to bring them to a stop. The elastic mechanical interaction of catapult and ship would be used to bring the *Solar Wind* to a gentle stop, catching it like a stretchy net catching a thrown ball.

"Well," said Kinet, "on the plus side, it means we won't need the landing legs. In a catapult landing, we'll be using the tailhook and the belly skids."

"That doesn't make me a whole lot happier," said Modulas. "The tailhook mechanism is attached to the same supports as the landing legs. I'm worried that it might have been damaged by the landing as well."

"All of the control boards and test equipment say that the tailhook is fine," said Teract.

"I know that. I just have a nasty feeling about it."

"Unless you can come up with more than that, command isn't going to want to hear it," said Stas. "If I'm reading the messages right, the repair office didn't get clearance from our research cluster before they drafted us for this mission. The head of small human research is furious."

© It's About Time

"There's not a whole lot to be done about that now," said Teract. "Kinet, take us home. Maximum boost."

"Aye aye," said Kinet. "Setting course for the *Wandering Star* and near-Earth space, maximum boost."

✳ ✳ ✳

The *Solar Wind* was moving like an Indy 500 race car with no brakes by the time it returned to near-Earth space. It was time for the crew to deploy the tailhook and try for a high-speed docking.

"Have I mentioned that I don't like this?" asked Modulas. His usual healthy glow had brightened so much that it was hard to look at him inside the dome of his suit.

Teract sighed. "Only about 400 times." She reached back with her tail and patted the shoulder of his suit. "It'll be all right."

"I'm with Modulas," said Stas. "We've got a damaged ship and we're at the max end of what the catapult can handle for landings. I think we should go into Earth orbit and bleed off some speed."

"Relax. We should be fine as long as the tailhook holds." Kinet's voice was even, but the muscles of all four of his shoulders were bunched with tension.

"That's exactly what I'm worried about," said Modulas. It was more of a mutter than a statement, and the others pretended not to hear it since there was nothing they could do about it.

The ship's speakers activated. "*Solar Wind*, you are cleared for landing on catapult two. You may begin when ready."

"I guess that's it then," said Teract. "Let's land this baby. Modulas, deploy the tailhook. Then you can take us in, Kinet."

"Deploying tailhook," said Modulas. He tapped a series of commands into the computer in his suit. "Tailhook deployed. All screens reading green."

"Beginning landing procedure," said Kinet. With careful blasts from the horizontal and vertical thrusters, he lined the ship up with the catapult. The distance to the *Wandering Star* shrank quickly. "Well, from here on out, we just have to hope that the tailhook makes a good grab."

"It had better," said Stas. "Otherwise, we're going to end up as a splintership pancake on the wall at the end of the catapult."

Seconds ticked by. In less than a minute, the *Wandering Star* went from looking like a shiny marble to a metallic basketball to a huge steel wall with a tiny hole in it, It was that hole they were aiming for and a moment later, it swallowed them whole.

© It's About Time

"Tailhook engaged!" said Kinet, as his board registered the connection. He let out the breath he hadn't realized he was holding.

They lurched forward against their straps as the catapult's spring system caught them and started slowing them down. It was a rough ride, but no one complained. They were going to be all right. Or at least it seemed that way.

"Red board!" Modulas shouted suddenly. "I have damage lights showing on two of the tailhook mounting brackets. They must have cracked in the landing on that asteroid. Now, the stress is breaking them up."

"Can you do anything about it?" asked Teract.

"Not a thing. We'll just have to wait and see. I don't know how it can hope to hold if we can't find some other way to slow down."

Teract turned to Kinet. "Any bright ideas?"

"Nothing. If I try to fire the breaking rockets inside the hull like this, I'll cook us all."

"I've got an idea," Stas said, suddenly. "Turn the ship!"

"Are you crazy?" asked Modulas. "That'll put even more strain on the tailhook!"

"No," breathed Teract. "She's right. Kinet, turn the ship as far as you can without disengaging the hook."

"All right. I'll do it, but I don't see how it's going to help."

"It'll put our tail and wing sections in contact with the walls," Stas explained. "The extra friction from that might slow us enough to keep the tailhook from giving way."

The tough hull metal of the splintership dragged along the equally tough catapult walls. The noise was horrendous. The sparks thrown up by the grinding metal surfaces blocked the aliens' view, but the added friction seemed to be working. The warning lights stayed on and it got very warm inside, but the tailhook held together and the four aliens were still alive when the ship finally screeched to rest just one meter from the wall at the end of the catapult.

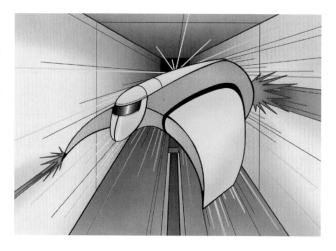

Several hours later, they inspected the ship and catapult. Long scrapes scarred the catapult tube where the splintership had rubbed against it, but that was just cosmetic damage. So were the burn marks on both ship and catapult from the heat produced by the surfaces rubbing against each other. The deep gashes on the skin of the *Solar Wind* were worse, but hull metal was tough stuff and their ship could be repaired. Between the hull damage, the tailhook, and the landing gear, the *Solar Wind* was going to be out of commission for several weeks. Still, the four aliens were just happy to be alive.

© It's About Time

Question

In Unit 3, you are learning about mechanical interactions (applied, drag, friction and elastic mechanical interactions). The aliens' adventures involve many mechanical interactions. Copy the table below onto your own paper. Then, from the stories, write down at least two examples of each type of interaction. Be sure to label the mechanical interaction type.

Type of Mechanical Interaction	Example
Drag	
Drag	
Applied	
Applied	
Friction	
Friction	
Elastic	
Elastic	

© It's About Time

© It's About Time

A Sticky Business

"Novas! I hate this," Kinet snarled as he pawed through a cabinet in the common area of the research cluster.

Stas stared. "What's wrong, Kinet? What are you looking for?"

"I'm trying to find a vial of information serum. I want a new knowledge shot."

"Which one are you looking for?"

Kinet shrugged all four of his shoulders. "I'm not really sure. That's part of the problem. That friction braking idea you came up with saved the *Solar Wind* when we crashed on the catapult deck. It was a really sweet maneuver. I've been thinking about it a lot lately."

"I'm just glad it worked," said Stas, rubbing her antennae together in relief. "That was one scary ride." She tilted her head to one side in question. "What does that have to do with a knowledge shot?"

"Well," said Kinet, "I realized that even though I kind of know how to use friction, I don't know that much about how it works. So, I'm trying to find a serum that'll have the explanation encoded in it. The problem is I don't know where to look. I mean, I could take the whole series for physics, but that's an awful lot of shots, and I really don't think I'll need all of it." He stopped digging around and turned his full attention to Stas. "How did you come to think of it?"

"Ahh…" Stas paused, looking embarrassed. Her features were made of hard chitin, so her face couldn't change expression, but Kinet knew her very well. Her emotion was clear from her posture and the way she held her antennae.

"All right," he said. "What's up? Come on, out with it."

"I've been reading the small human learning books," said Stas. "I was reading them to get a better idea of how the humans teach their young. Then I found out that they had explanations I didn't really understand. Things that I always thought I knew."

Kinet swiveled his ears forward in interest. "I don't get it."

"It's like you and this friction thing. I'm sure that somewhere in your pilot and researcher shots you got a basic explanation of how to use friction in your job. But clearly no one bothered to put in an explanation of how it works. The shots are nice, but they leave gaps. They give you information, but they don't teach you how to think. If you don't really understand something, it's hard to make use of it. You have to be able to construct and evaluate explanations for things to really say that you understand them."

© It's About Time

Kinet was shocked by the bitterness in her voice. "Do you think the human way is better?"

"I don't know," she said, gently shaking her head in a human gesture she'd learned. "From what I've seen, the human way doesn't always encourage thinking either. Some of their classes are just about memorizing things. It's knowledge without understanding, like our shots, only slower. In other books and classes, they really encourage their young to learn."

"And you learned something about friction from one of these *books*?" The word tasted funny in Kinet's mouth.

"Yes."

After a long pause, Kinet spoke again. "Will you show me?"

"Sure." Stas led him back to the pod lounge where she pulled out a portable reader. "I've been loading every human text I could find into this. Let me just do a search for friction in the basic science section, and note that I want to check things I've already looked at…"

She entered her search commands. After a brief delay, the information popped up. Kinet and Stas read together for a while.

"So," said Kinet, "the humans still don't have a complete understanding of exactly how friction works, but they do have some good models of its behavior."

"Yes," agreed Stas, "this thing they call the 'bump model' explains it pretty well. When two surfaces rub against each other, tiny bumps on the surfaces interact by pushing against each other."

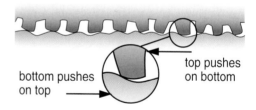

side view of sandpaper

bottom pushes on top

top pushes on bottom

"Each of these little pushes is a force that works against the smooth passage of the surfaces one past the other." Kinet smiled, then frowned. The explanation was simple, it made sense, and it hadn't been included in his pilot shots. He added his own ideas to this explanation. "So, when our ship was sliding along the walls in the crash, the little bumps worked to slow us down."

"And generated a great deal of heat in the bargain," added Stas. "There was a lot of energy transfer involved."

"Sure," said Kinet. "That's where most of our forward motion went. The rubbing surfaces converted the motion energy into heat energy. I remember the hull was so hot it almost glowed." He paused, a thoughtful look on his face. "You know, I do have something about this tucked away in my head. It came with my pilot shots."

"What's that?"

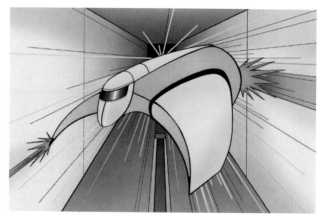

© It's About Time

"Drag. Whenever a ship enters atmosphere, it heats up from friction."

"How can that be?" asked Stas. "Friction with what?"

"Air," said Kinet. "The ship interacts with the air as it flies through an atmosphere." Stas looked confused and started to open her mouth, but Kinet held up a hand. "Wait. Let me try to explain. You've taken baths, right?"

"Of course!"

"Well, when you push your hand through the water, it moves more slowly than it would in the air, and you have to push harder to make it go the same speed, right?"

"Sure," said Stas, "but that's because water has some density to it."

"So does air," replied Kinet. "I know that it doesn't seem that way, but it's really pretty significant. Especially when you compare it to the vacuum of space. In order to pass through all that air, you have to push it out of the way, and it drags along the sides of the ship. I bet it works the same way that the bumps work in the description in that book, a sort of friction interaction between the air and the ship. I wonder…" He scratched behind his left ear. "Do you suppose that somewhere in our database we have a better explanation?"

"I really don't know," replied Stas. "I'd like to think so. After all, we've had a couple thousand years more than the humans have to discover how things work."

Kinet sighed. "I was afraid you were going to say that."

"Why afraid?" Stas cocked her head to one side in puzzlement.

"Because now I have to go back out and dig through the cabinet again. I want to know more. I really hope I don't have to take the whole series of physics shots."

"Before you do that, why don't you try some more reading," suggested Stas. "Maybe the humans have learned more since they wrote this book." She tapped her screen with one long and elegant finger. "It might take a little while longer to do it this way, but it's sure less painful than the shots."

Questions

1. Write a few sentences describing the alien's bump model for accounting for friction.

2. Draw an energy diagram describing the friction interaction between the splintership in the story and the walls of the mother ship.

© It's About Time

© It's About Time

A Night for Newton

It was dark when Kinet woke up and checked the clock. It was almost three in the morning. He'd been having a dream about crossing one of Earth's deserts. His own planet didn't have any of the harsh places. Instead, Anima's land surfaces were all densely forested.

Kinet tried to go back to sleep, but the dry and parched images of his dream kept haunting him, making him thirsty. He decided that there was no way he could get back to sleep without having a drink of water first. So, reluctantly, he pulled on a robe and stumbled out of his room.

He'd gone less than two meters when he noticed Stas. The insectoid alien lay on her back on the floor of the lounge, reading. She clutched the small electronic reader in one hand. Her other hand was tucked under her head as a cushion. The long, gripping toes of her feet were wrapped around a chair arm.

"Do you know what time it?" asked Kinet, his voice rough with the edges of sleep.

"Sure, It's about two-forty-five."

"Then why are you still up?"

"I was planning on going to sleep hours ago," admitted Stas. "But I was skimming through a physics text book and I ran into something fascinating."

"In a physics text?" Kinet's voice was incredulous. "I know that science can be fun and all that, but I have trouble imagining something like that would justify staying up this late."

Stas clicked her mandibles together in her species' equivalent of laughter. "I can understand your surprise. I don't think I'd still be awake if I were just reading straight science." She paused for a moment, and tilted her head to one side. "Though you never can tell. No. What I find so interesting is actually a bit of history."

© It's About Time

It was Kinet's turn to laugh. "I think I begin to see. Your entire species seems to go nuts for history. What did you find this time?" Now totally awake, he plopped down on the couch. He figured that he might as well find out what was keeping Stas up before he tried to go back to bed.

"I was glancing through the section on motion and I hit a little piece of biography. That struck me as odd, so I read it. Here," Stas handed him the reader, "take a look."

Kinet took the device and began to read aloud. 'Sir Isaac Newton (1642–1727). Born prematurely into a farm family in Woolsthorpe England. Father died before he was born. Left to live with his grandmother when his mother remarried. Didn't do well at farming, but showed promise as an academic. Went off to Trinity College, Cambridge, and took a degree without distinction. Sent home in 1665 when plague closed the university.' I don't know, Stas. It doesn't sound very interesting…"

"It's not his comings and goings that are so interesting." Stas said impatiently. "It's what he did. He proposed the three laws that describe the relationship between forces and motion: inertia, force, and action-reaction."

"Oh! Like Tantaris on Anima." Kinet referred to one of the early scientists of his own world's history.

"More like the great Ectovorian philosopher Betalma. Because, like her, Newton didn't just come up with the rules of force and motion, he also did fundamental work in mathematics, chemistry, gravitation, astronomy, and optics."

"Really? It seems almost too much of a coincidence. What sorts of things did he work on?"

"In optics, he came up with the idea of white light being composed of colored light. He did all sorts of things with refraction, and the particle theory of light. He also formulated much of the theory of calculus. All just like Betalma."

"All that?" asked Kinet. "Really?"

"Definitely all that, and much more besides. I'm still looking for information on Newton's chemical and gravitational work. Betalma was responsible for the theory of chemical force on Ectovoria. I've been trying to find out whether Newton did anything like that. Unfortunately, all the stuff I've read so far in this book has focused on the fact that Newton studied something called alchemy.'"

"Is that bad?" asked Kinet.

"Apparently, alchemists were people who tried to change common basic metals like iron into more valuable metals like silver or gold. They were supposed to have taken dangerous risks for dubious ends. Their experiments were rarely guided by a theory or a set of ideas supported by evidence. Many of the scientists of Newton's time disapproved of them. Some of the stuff I've been reading suggests that Newton conducted alchemy research of his own and managed to poison himself with mercury in the process."

© It's About Time

"Nasty stuff," said Kinet. "Mercury can destroy the brain."

"It certainly can. Here on Earth, mercury was used by people who made hats. Because of mercury poisoning, they went insane so often that they had a sickness named after them. It was called 'Mad Hatter's Disease.' At the time, they didn't know that the mercury was causing it."

"And Newton got it?" asked Kinet.

"That's what the historians believe, but he recovered."

"Wow! He must have been an incredible individual."

"He sure was. Right up there with Betalma. The similarities are amazing."

"I don't know how amazing it is," said Kinet. "Science is just a process of exploration. No matter where you go, the rules are the same. Maybe some people are just better at finding out what those rules are. If that's true, is it such a surprise that really brilliant individuals from two different species could discover so many of the same things? Weren't they both building on the work of those who came before them? At least in part?"

"Funny you should mention that," said Stas. "Newton was one thing that Betalma was not. He was modest. When asked about his genius, he once said 'If I have seen further, it is by standing upon the shoulders of giants.'"

Kinet laughed. "Now it makes sense."

"What makes sense?"

"As a space pilot, I got interested in Earth's early space program and the race to the Moon. One of the astronauts on the first Moon launch said something about standing on the shoulders of Newton."

The sound of Stas' mandibles clicking mingled with Kinet's laughter.

© It's About Time

Questions

1. After reading this story, what else would you like to find out about Isaac Newton? Name at least one thing.

   (Many Internet sites have information about Newton – take a look!)

   http://www.lib.cam.ac.uk/MSS/Newton.html,

   http://landau1.phys.virginia.edu/classes/109/lectures/newton.html

   http://www.physics.ucla.edu/class/85HC_Gruner/bios/newton.html

   http://www.pbs.org/wnet/hawking/cosmostar/html/cstars_newt.html

2. Isaac Newton proposed the relationship between forces and motion. Based on what you have learned so far, answer the following questions:

   a. If an object is speeding up, what can you say about the forces being exerted on the object?

   b. If an object is slowing down, what can you say about the forces being exerted on the object?

   c. If an object has a constant speed, what can you say about the forces being exerted on the object?

© It's About Time

Not That Newton

Stas was napping in the lounge when Kinet burst in. He was clutching a grocery bag and grinning like a maniac. "You didn't tell me that Newton was also a great cook." Kinet waved the bag at her.

Stas blinked at him. "What in the galaxy are you talking about?"

"The scientist Newton, remember? You told me all about him a couple of nights ago."

"Newton wasn't a cook."

"Then what do you say to these?" Kinet dumped out his bag onto the small table in front of Stas' couch. A stream of snack food cascaded out. Snack food had been Kinet's latest discovery about Earth. Trade Confederation ships carried only relatively bland, nutritionally balanced foodstuffs. Earth, on the other hand, seemed to be practically overrun with things that tasted wonderful but might not fit into everyone's idea of a balanced diet. Kinet loved the stuff. He pawed through the pile and produced a cookie package. The label read 'Fig Newtons.™'

"These things are great," he said. "That Newton was one heck of a cook."

Stas clicked her mandibles together in the Ectovorian equivalent of laughter. "I'm afraid Isaac Newton didn't have anything to do with your snacks, Kinet."

"What?" asked Kinet. "His name's right on the package. Besides, didn't you say something about Newton making an important discovery with apples?" Kinet started digging through the pile again. This time, he pulled out a package of Apple Newtons. "I don't know if these are as good as the fig kind, but if so, I'm sold."

Stas' mandibles were clicking so fast they sounded like raindrops on a shed roof. "Oh, Kinet. Newton *did* make a discovery that involved apples, but it was *scientific*, not culinary. One day, as folklore goes, he was sitting on an estate just thinking, when he saw an apple drop from a tree. It led him to wonder if the force that caused that apple to fall to Earth was the same as the force that governed the motion of the moon."

Kinet looked incredulous. "From falling fruit, he figured out gravity? That doesn't sound right. I think cookies are a much more likely fruit discovery. Besides, the name's the same."

© It's About Time

"It wasn't quite as simple as that," said Stas. "He didn't immediately figure out the principles of gravity from seeing an apple fall. It made him think about the subject of interaction at a distance. He figured that for the apple to fall, it must somehow be interacting with the Earth. After a lot of hard thought and mathematics, he came up with the basic principle of gravitation. It was very controversial at the time too. A lot of people disagreed with him"

"I should think so," said Kinet. "Seeing the Moon in a ripe apple. That sounds more like poetry than science."

"Actually, it was the idea that force could act at a distance that was the problem. Most people didn't believe it."

"Really? That seems surprising now." Kinet shrugged both sets of shoulders. "I suppose it's no odder than my notion that a physicist could cook. I mean, how likely is that?"

"I've known some scientists who were excellent cooks," Stas said indignantly. "You're teasing me, aren't you?"

Kinet shook his head solemnly, but then started giggling. He couldn't help himself. "This whole thing was a joke, wasn't it?" demanded Stas.

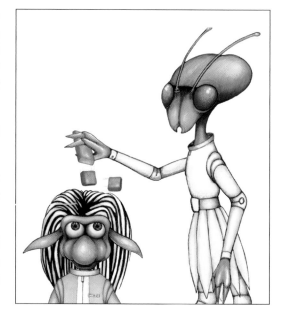

"Well, the cookies are actually pretty good. But the rest... Yeah, and you fell for it hook, line, and sinker." He began to giggle again and the giggles turned into outright guffaws. A few minutes later, he was rolling on the floor laughing.

"I don't think it was that funny," Stas said but her mandibles were clicking softly.

"That's only because you couldn't see your face." Kinet's words came between gasps for breath.

A twinkle lit Stas' eye. "You know, Kinet, I don't think you're conducting yourself with sufficient gravity. Try this." She dumped the package of apple cookies over his head.

Questions

1. What was very controversial about Newton's ideas on gravity?

2. Write a short paragraph that describes what you have learned about gravitational interactions.

© It's About Time

That Sinking Feeling

"I sure hope they go for it," said Kinet.

"Me too," said Stas, snapping her mandibles sharply together once to signal her agreement. "We could really use a break from studying the small humans. We've been assigned to them since we first made contact with the humans eight months ago. But I'm really not so sure about this camping thing."

"It'll be fun," said Kinet. "Besides, Earth mountain camping isn't something any of the crew of the *Wandering Star* has done yet. So Teract should be able to sell it to Cluster Command as research as well as a vacation."

"I understand that, but couldn't we have gone to Paris for art appreciation and shopping instead?"

"We could," interjected Modulas, "if our research Cluster director hadn't already snagged the Paris cultural exchange as his vacation spot. We're still a very junior research pod in a great big ship. We'll be lucky if we get a vacation at all."

"Too right," said Kinet. "Come on, let's go down to the hangar and grab the gear we're going to need from the *Solar Wind*."

✳ ✳ ✳

"Whoof," said Kinet, as he tried to lift a package from its rack in the hold of the *Solar Wind*. "This is really heavy." All four of his hands were firmly grasping a slightly wrinkled cube of neon green plastic. On the side opposite him, a small metal tank was strapped to the cube with a slender tube running from its valve to a similar valve on the cube. The whole package was about 40 centimeters on each side. "Hey Teract, could you give me a hand?"

"Sure," said the lizardlike alien, as she got a grip on the metal of the tank. Together, they were able to lift it onto a wheeled cart with the rest of their equipment for the trip. This included several small cases of food, their personal clothes, and two much smaller cube and tank arrangements. All of the items were part of the standard survival equipment that belonged to their splintership.

"What does that thing weigh?" asked Kinet.

"Over a hundred kilos. That's almost a thousand Newtons at Earth's gravity," replied Modulas, using the electronics in his suit to use the cart's built-in scale. "That's almost half of our total luggage allowance for this outing, but we can't make the trip without it."

© It's About Time

A few hours later, a space-to-surface shuttle from the *Wandering Star* dropped them off in a parking lot at Ville de Waterton on the Canadian side of Glacier-Waterton International Peace Park. From there, they would travel the length of the lake to Goat Haunt on the American side, where they would camp for four days. The late afternoon Sun was already dipping toward the mountains as they unloaded their gear.

As soon as the shuttle took off, Teract and Stas each took one of the smaller plastic cubes and headed across the parking lot to the dock. Modulas picked up their personal packs and followed after, leaving Kinet alone with the big green cube. He eyed it dubiously. He knew from his earlier experience that he couldn't carry it all by himself, but he had an idea.

He quickly knelt and opened the valves. With a hiss, the cube began to expand and unfold as the compressed air in the tank rushed into the inflatable boat. Within five minutes, the cube had grown into a boat about two meters wide and six meters long. Unlike a similar device of Earth manufacture, this was no mere raft. It was an actual cabin cruiser with built-in chairs and an enclosed pilot's cabin complete with steering wheel. It was also much more rigid than an Earth-made inflatable raft, almost as solid to the touch as a wooden or fiberglass boat.

"There," he said to himself, "that should make it easier." He bent and slid his hands under the back end of the boat and lifted. To his surprise, it seemed just as heavy as it had before he inflated it. But now it was even more awkward to move. He was still trying to figure that out when Teract and Modulas returned. They'd left Stas to watch the rest of their gear.

"Why did you inflate it?" cried Teract, as she walked around the side of the boat.

"I figured it would be lighter this way," mumbled Kinet.

"How so?" asked Teract. "All you did was transfer the air that was stored in the tank into the boat. It's all still there."

"Yeah," said Kinet, "but…"

Teract shook her head. "We can talk about it later, over dinner. Right now, we need to get the boat down to the water and get moving if we want to make it to the other end of the lake before nightfall." She went to the bow and got a grip. "Modulas, why don't you guide us and tell us where to go. With the boat inflated like this, I'd rather you didn't try to carry it. Your suit's grippers might tear the plastic."

✳ ✳ ✳

An hour later, with the Sun half hidden by the peaks, they were nearing the south end of the lake. Because it had taken them a long time to get the boat into the water and loaded, Teract, who was driving, was pushing the little engine to the maximum.

"It's a good thing we're almost there," said Stas, from where she sat in the bow. She stood up to peer toward shore only a few hundred feet away. "I'm hungry."

"Me too," agreed Teract. "I'm sure looking forward to …" She didn't get to finish the sentence because the boat hit the jagged end of a log that was floating just below the surface.

© It's About Time

The impact almost stopped the boat dead in the water. Kinet and Teract were thrown out of their seats and into the boat's windshield. Fortunately, it was made of a thick flexible plastic and neither was seriously injured, though they were both shaken. Modulas, sitting on the floor in the back to balance out Stas in the front, slid into the back of Kinet's seat. But the insect girl was not so lucky. While the boat stopped moving, Stas did not. She sailed over the front railing to land with a splash in the icy water of the mountain lake.

A moment later, Teract kicked off her boots and followed after Stas. The blue alien came from an ocean planet, and her people lived as much in the water as out of it. Even so, she found the deep waters of the lake almost unbearably cold, fueled as they were by melting snow. Sliding swiftly through the dark and frigid waters, she went straight to the place where Stas had gone in. She was relieved to find the insectoid alien floating with her head above water.

"Are you all right?" she asked.

"Y-y-yes," stammered Stas, bobbing her head in the affirmative as she had learned from the humans. "J-j-just c-cold. I couldn't sink if I wanted to, but I c-can't say the same for the boat."

She jerked her chin in the direction of the little craft, and Teract turned her gaze that way. The hull was visibly sagging, and an ominous trail of bubbles rose from where the boat remained hung up on the log. Clearly, it was leaking air.

"I c-can get to shore ok-kay," said Stas. "It's not that far. W-why don't you help the others. I don't know how well K-kinet can swim, and I'm worried about Modulas."

"Are you sure you're all right?" asked Teract.

Stas nodded again and started paddling for shore. Teract swam quickly back to the boat and pulled herself over the rail, which was now much closer to the water. She didn't like what she found there at all. Kinet was clinging to the highest point in the boat and pointing frantically to the back where Modulas stood. Underneath the heavy metal boots of the alien's suit, the deflating hull was severely damaged.

© It's About Time

"Oh gralf!" she whispered to herself. "Modulas, do you have a flotation device tucked away in that suit somewhere?" Suddenly, the boat shifted under them and water started to pour in over the back rail and rise around his feet.

"No," he said, his voice only a little panicky. "As long as nothing shorts out, and the lake isn't too deep, I should be all right. Worry about Kinet, he can't swim." The boat shifted again, and Modulas fell over the side, sinking like a rock.

There was nothing Teract could do about that for the moment. Even if she'd been able to get a hold of him there was no way she could have held the suit above water. She turned her attention to Kinet who was checking the straps on his life vest.

"Come on," she said, taking his hand. "We need to get off this boat before it goes under completely."

A moment later, Teract was towing him toward shore. His grip was tight enough to hurt, and his body was practically rigid, but she soon had him at the edge of the water. Because it was a mountain lake, the change from deep to shallow was abrupt and close to shore. In the knee-deep water, she helped Kinet to his feet, and gave him a gentle shove toward dry land. With his fur soaked, he weighed a ton, and had trouble staying upright, but Stas came back into the water and gave him a hand. As soon as she did, Teract turned and headed back out to the boat. Just as she reached it, it came loose of the log and sank into the dark waters.

The light was fading fast and, before the boat had sunk 10 meters, it vanished from sight. Taking a deep breath, Teract dived after it. She didn't really care about the boat just then, but it marked the place where Modulas had gone under. For more than a minute, she swam straight down, propelled by her webbed feet and powerful sweeps of her arms. Soon, she was beyond reach of all light, but had not yet found the bottom. She began to worry that this part of the lake would be too deep for her. She had read that it reached 152 meters in some places, the deepest lake in the Rockies. Just when she thought she would have to turn back, her hands touched the bare rock bottom. Flipping over, she peered into the black waters around her.

She didn't know what she was looking for, since she couldn't see anything. But she knew that even a Sircosian couldn't stay at depths like this for long, and she needed to find Modulas. Suddenly, she saw a pinpoint of light off to her left. Swimming quickly in the direction of the light, she found Modulas, walking along the bottom toward the shore. She was nearly out of air, but she swam around and into the beam of his lights to avoid startling him. She signaled him to see if he was all right. When he held his hand up in the Earth gesture for okay, she nodded once and headed for the surface.

Now that she knew her whole crew was going to be all right, she really started to feel the cold, and the lack of oxygen. By the time she made it to the surface, she was gasping and half frozen. The swim back to shore seemed to take forever, and when she got there, she needed Stas' help to stand up and stagger out of the water.

"Is Modulas here yet?" she asked, in a voice barely above a whisper.

"No," said Stas. "We thought you ..."

© It's About Time

Kinet interrupted her. "I see a light under the water! There's a trail of bubbles! I think… Yes, it's Modulas!"

"Great," said Teract. "Now I can collapse." A moment later, she sagged in Stas' arms.

Modulas hit the shallows, and with water still beading off the dome of his suit, he scooped Teract up and carried her ashore.

"Now," said Kinet, "if we can just get a fire going, we'll be all right."

Questions

1. When Kinet released the compressed air from the air tanks into the inflatable boat, the boat and air tank combination did not become less heavy. Why?

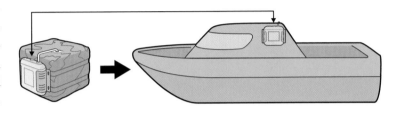

2. In Interactions in Physical Science, you have studied many types of interactions, such as electric circuit, magnetic, light, applied, elastic, friction and drag interactions. Copy the table below onto your own paper, and then find 10 examples of different types of interactions in *That Sinking Feeling*. Cite the examples in your table. Try to find examples of different kinds of interactions. For instance, don't cite only applied interactions, or only light interactions.

	Interaction Type	**Interacting Objects and Description**
1.	Drag	There was a drag interaction between the boat and the water when the boat sank into the lake.
2.		
3.		
4.		
5.		
6.		
7.		
8.		
9.		
10.		

© It's About Time

© It's About Time

Stranded

Modulas lowered Teract gently to the ground beside the empty firepit. She was an even paler blue than usual, shivering and only half-conscious. After their boat hit a log and sank she had spent considerable time in the water, making sure everyone else made it to shore safely. Even though she was from an ocean planet, and as much at home in the water as out of it, Waterton Lake was fed by melting snow and ice, and it was much colder than the seas her people normally swam.

"Brr," shivered Kinet, hugging himself with all four arms. He too had ended up in the water, and his fur hadn't helped him at all. Quite the contrary. While the thick coat that completely covered his body normally protected him from extremes of temperature, it was now saturated by the near freezing water. Not only was he cold, he also weighed a great deal more than usual. Everywhere he went, he left a dripping trail. He tried taking off his uniform and shaking himself off, but it didn't help much.

"W-we need s-some wood," stuttered Stas. "To g-get a fire going." The water had quickly run off the thick chitin of her shell and left her dry. Like most insects, she needed to stay warm, and with the Sun going down, the high mountain campsite was getting colder fast. Her antennae were vibrating gently from the chill.

"I'll get started," said Modulas.

Though he was actually the most fragile of the four aliens, his suit provided him with a comfortable environment even in extreme temperatures. Hitting a switch on the control panel of his suit, he activated his exterior lights and headed into the nearby woods. The others huddled together while they waited for his return. He was gone only a few minutes, but it seemed a very long time to those left behind.

"Here," said Modulas, dropping a small pile of dead branches and twigs into the pit. "This'll make a good start. I'll be right back."

© It's About Time

"W-wait," said Stas. "We don't have any matches or anything to get the fire going. Do you?"

He shook his head. Looking down at the quiet and still form of Teract, he felt real fear take hold of him and the faint luminescence of his skin bloomed brighter with agitation. "We *have* to start a fire for Teract, and we'll need more wood for that. See what you can think of while I get it."

As Modulas moved through the darkening woods, nothing came to mind and he grew more and more scared. What if they *couldn't* make a fire? Frightened at the way things were going, he smacked his small fist against the control panel of his suit. The impact hurt, a lot, and he instantly knew that he would bruise. The injury shocked him out of his near panic. His race's fragility was the reason they had to wear their suits and the pain was a warning to control himself. Still, when he returned to the camp, his skin was blazing as bright as any flame. He dumped his burden next to Stas who was carefully piling twigs and bark into a little cone.

"Any ideas?" he asked, and was enormously relieved when Kinet nodded yes.

"The only reason I thought of it was because we did so much work on your suit after the *Solar Wind* fell on you. If we can get at the wires for your light system, we can create a short circuit and maybe use the resulting hot wires to start some of this bark on fire." Stas held a piece of kindling up in the air. "Stas is going to do all the outside work because my wet fur would put any fire I started right out again."

A few minutes later, Stas had removed the watertight cover of Modulas' left headlight, and pulled out the wires. Meanwhile, inside the suit, Modulas bypassed the safety system that would normally disconnect the electricity to the lights in case of a short.

"Ready?" asked Stas.

"As ready as I'm going to get," replied Modulas. "Try and make it quick. This is going to seriously drain my batteries, and you really don't want to have to carry my suit out of here."

Modulas hoped he sounded calm, but the thought really worried him. His powered suit was the only thing that kept his fragile body protected from the rigors of the outside world. If its power failed, he'd be in real trouble. He crossed his fingers in a gesture he'd learned from the humans. "Do it."

Stas touched the wires together and placed them against a bit of dried bark when they started to heat up. The wires quickly began to glow red and the piece of kindling began to smoke where it touched them. The instant it burst into flame, Stas released her grip on the wires, undoing the short and ending the energy drain on Modulas suit. Moving carefully so she wouldn't snuff the tiny flame, she inserted it into the little cone of twigs and bark.

© It's About Time

As the fire took hold and gradually began to grow, Stas steadily added larger and larger pieces of wood, until they had a good-sized fire going. Soon they were able to huddle around it for warmth, and Teract was restored to full consciousness. They told her what had happened while Stas reassembled Modulas' light system.

"That's kind of funny," Teract said, when they were done.

"What?" asked Kinet. He was practically standing in the fire now, and the water in his fur was steaming away into the night air. He wished he could stand closer, but he wanted to get warm, not catch on fire. That's exactly what would happen if he came into actual contact with the flames instead of just soaking up the heat that radiated from the fire.

"The *Wandering Star* uses fusion reactors to produce heat energy which in turn drives turbines to make electrical energy. Then that electricity is converted to chemical energy for storage in Modulas' batteries, and back to electricity for his lights, which you shorted to convert it back to heat. So really, that fire was born from the same energy that powers the *Wandering Star*. It might have been transformed many times, but it's still the same energy."

Stas nodded. "For that matter, the fusion that drives the *Wandering Star* is the same process that fuels the Sun. And this fire, not just the initial heat we used to start it, ultimately comes from the Sun."

"Huh?" said Kinet.

"Well," replied Stas, "trees eat sunlight in much the same way we eat food. So the tree is converting the light from the Sun's fusion into chemical energy that it stores as part of its structure." She tapped a piece of wood. "Now, by burning the wood, we're converting it back to heat energy. It's all a big circle."

© It's About Time

"I suppose you're right," said Kinet, "but it's hard to understand. How do you suppose it all works?" He shook his head. "No, don't try to explain. I just want to sit here, enjoy the warmth for a while, and think about what we do next. This will keep us warm until morning, but all of our food, and our communications equipment is on the bottom of the lake. We're still in trouble."

Questions

1. Find at least two examples of thermal interactions in *Stranded*. Describe each interaction, and draw an energy diagram for each one.

2. Find at least two examples of infrared interactions in *Stranded*. Describe each interaction, and draw an energy diagram for each one.

3. Before the fire was built, Kinet was dripping wet. As he warmed himself by the fire, the water in Kinet's fur evaporated. Describe the interaction that changes the state of the water from liquid to gas, and draw an energy diagram for the interaction.

© It's About Time

Signal Strength

Kinet was up with the dawn. Since he normally just curled into a ball to sleep, he'd had a pretty good night, despite the boat wreck that had left them stranded. Once the water had evaporated out of his coat and he'd gotten dry, the fur had kept him warm and allowed him to sleep quite comfortably. After stretching all six of his limbs until they creaked and yawning enormously, he rose to his feet.

The four of them had slept in a circle around the fire. The fire lay in the low pit the park service had dug for bonfires, and their meager blaze seemed tiny in comparison to the space given it. This was especially true since it had mostly burned out in the hours before sunrise, leaving only a few coals. When Kinet raided their little woodpile to rebuild the fire, he was surprised to find a thin layer of frost covering the ground beyond the pit. It *had* gotten cold.

As he blew on the coals to get them to blaze up and ignite the fresh wood, he heard a stirring from behind and to his right where Teract had been sleeping.

"Wha's up?" mumbled Teract.

He pointed at the coals with one hand while still keeping the other three busy feeding twigs into the flames. "Just getting this going again." His tone was quiet, almost whispering. He didn't want to wake the others. "How are you doing?"

"Stiff," said Teract, "but otherwise pretty good."

Kinet wasn't surprised. Teract was tough and athletic, and probably in better shape than any other member of her crew.

"So what do we do next?" she asked.

"Don't ask me," replied Kinet. "I just pilot starships. You're the brains of this outfit."

"The brains who drove our boat into a tree you mean." There was a bitter undertone to her words that made Kinet look up sharply.

"No kicking yourself," said Kinet. "That's the first rule of crash landings." He grinned. "That's the passengers' job. Seriously, it could have happened to any of us. There was just no way to see that tree."

© It's About Time

"He's right," said Stas, who now unfolding herself like a flower opening to the sun. "Besides, nobody was hurt. All we lost was some equipment."

"Including all of our food," said Teract, "and our radio. There's no way we can call for help."

"I wouldn't worry about it too much," said Kinet. "We've only got four more days of vacation, and there's no way Cluster Command is going to extend that. If we aren't there for the shuttle pickup on Wednesday, they'll come looking. Cluster Command wouldn't want us to miss any work."

"Four days is a long time," replied Teract. "We should have stayed on the *Wandering Star* and kept studying the small humans. We could get mighty hungry in four days."

"True," said Kinet, nodding, "but some other camper might come along before that. Or we might be able to find some food. Or, if all else fails, there's always the time honored tradition of eating our leader." He smiled evilly. "Would you rather be baked or roasted?"

This time, Teract laughed too. "All right. It's not a total disaster. I do wish there was something we could do. I just hate to sit here passively waiting for rescue."

"How about Modulas' suit?" asked Stas. "I know he has some sort of radio in there so his computer can talk to the computers on the ship."

"No good," said Modulas, who was awake now. "It's designed for local area networking only. The range is about a hundred meters. Even if we built a booster antenna, I don't think we could get it to reach more than a kilometer. That wouldn't get us even halfway across the lake. Besides, it's tuned for ship frequencies. I don't think any of the natives use the same range."

"Oh," said Stas. "Well, there goes one idea."

"Too bad," said Teract. "It would have saved me a swim."

"What?" asked Kinet. "You're not thinking about going back in the lake, are you?"

"I am. A bunch of the food is in watertight packages. If I can find it, we'll have enough to get by. I just wish I knew exactly where the boat went down so I wouldn't have to search around too much. The water is really deep and I can't stay on the bottom for very long."

"Maybe I can help you there," said Modulas. "My suit keeps a record of all the actions it takes; how many steps, how long the steps are, what direction. I can run back the data and see where I sank. That should get you pretty close."

"That's great!" said Teract. "I don't suppose you can walk back out there, can you?"

© It's About Time

"I'm afraid not," replied Modulas. "My onboard air supply was just barely enough to get me to shore. There's no way it can take me out and back again. In fact, I haven't even recharged it, because I don't want to use any more energy than I absolutely have to. I should have enough to get me through till we leave, but.... I'd planned on recharging from the boat's systems as I needed to."

"Right," said Teract. "Stas, you're in charge of getting the fire built up. It's still a little cold for you out here. Kinet, you get more wood. Modulas, you and I are going down to the lake."

✳ ✳ ✳

Enough sunlight filtered down through the clear mountain waters to let Teract dimly see her way around. It took half a dozen dives and she had to return to the campfire twice to warm up, but she finally found the boat. After four more dives, she assembled all of their gear in a heap. There was still one problem. While it had been relatively simple to move the four packs and the two deflated shelters when she had the bottom of the lake to push against, she didn't think she'd be able to lift any of them to the surface just by swimming. While she could take the gear out of the packs and bring it up in

smaller trips, there was no way for her to do that with the inflatable shelters. Also, she didn't know if she had enough stamina to make that many more dives.

Teract was already hungry when they hit the log the night before. After working for what seemed like hours in the icy water, she felt as if she were on the brink of starvation. She was also almost out of air, and desperately needed to warm up, so the solution would have to wait for her next dive. Later, as she sat shivering beside the fire, she related the problem to the others.

"What if you inflated one of the shelters?" asked Stas. "It should float to the surface."

"That's not going to make it any lighter," said Kinet. "I discovered that when I tried to inflate the boat to make it easier to carry. The weight of the system is going to stay the same, whether the air is in the tanks or inflating the shelters."

© It's About Time

"Sure," said Stas, "but weight isn't the only factor. Look, when we hit that log, it put a rip in the boat. That let the air out, so the boat, minus the weight of the air was actually lighter. When it was full of air it floated, and when it was empty, it sank. It should work the same way for the shelters. It's like an air mattress; inflated and heavier, it floats; empty and lighter, it sinks."

Kinet held up his hands in surrender. "I believe you, but don't try to explain any more. You're giving me a headache."

The idea worked, and Teract was able to use the shelters as floats to raise the packs. Later, as they sat around the fire toasting bits of dehydrated Binth for supper, Kinet found himself fascinated by the way bits of ash and cinders rose gently into the air above the flame.

"Hey," he said, sitting up suddenly, "I have an idea."

"Be careful," said Modulas, "you wouldn't want to hurt yourself."

"Oh yeah," said Kinet. "Same to you, buddy. You're going to have to apologize for that when you hear how brilliant I am."

"Sure," said Modulas. "Right. Go for it."

"Floating the shelters is what started me thinking. They reminded me of balloons. Sitting here watching the sparks rise, I wondered if we couldn't use one of the shelters as a hot air balloon, to make a signal. They're fireproof, and we could take the air tank off to make it even lighter. If we tied some fishing line to one so we wouldn't lose it, and held it over the fire to fill it with hot air, could we use it as a signal?"

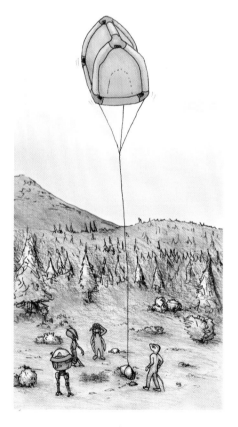

"I don't know," said Teract, "but it's worth a try."

A few hours later, when the rangers arrived, wondering what the big silver balloon was for, they were very surprised to find themselves rescuing a group of aliens. As the big ranger boat pulled away from shore, Teract and Stas congratulated Kinet for his idea and Modulas apologized for making fun of him. Kinet said "I told you so" to Modulas and stuck his tongue out at the other alien.

© It's About Time

Questions

Read the following portion of the second paragraph of *Signal Strength*, then answer the questions below.

"The four of them had slept in a circle around the campfire. A small fire still burned in the pit the park service had dug for bonfires. The remains of the campfire seemed tiny in the big fire pit. It had mostly burned out in the hours before sunrise, leaving only a few red coals and a weak flame. When Kinet raided their little woodpile to rebuild the fire, he was surprised to find a thin layer of frost covering the ground beyond the pit. It was evidence that it had really gotten cold overnight in the mountains."

1. Think of the burning wood as a system. Is mass conserved in this system? Write your reasoning.

2. While the wood was burning, the ground under the fire and around it must have warmed up. Draw an energy diagram that describes how the ground got warm. Name the kind of interaction that took place.

3. Later in *Signal Strength*, Teract fetched the alien's supplies from the bottom of the lake. During her dives, she got very cold. Draw an energy diagram that describes how Teract got cold. Name the kind of interaction that took place.

4. Teract sat in front of the fire to warm up between dives. Draw an energy diagram that describes how Teract got warmer. Name the kind of interaction that took place.

5. When Teract is interacting with the fire, is any energy created from nothing, or destroyed into nothing? Write your reasoning.

© It's About Time

© It's About Time

Broken Dreamers

"Wow, I'm beat," said Stas, tossing her uniform jacket on the lounge couch and sitting down beside it.

"Me too," agreed Kinet, folding himself up to sit cross-legged on the floor.

"Yeah," said Modulas. "I thought vacations were supposed to be restful." His suit provided him with all the chair he would ever need.

"Most vacations don't start with a shipwreck," said Teract. She still felt badly about the accident that had sunk their boat and stranded them in one of Glacier Park's wilderness areas. "I should go report in. We need to get back to studying the small humans, and I need to request a replacement boat for the one I lost."

"Sit!" commanded Stas. "We just got home. You're not going anywhere until you've had some rest."

"All right." Teract flopped down next to Stas. "You convinced me. I'm totally toasted, and I really don't want to face the chewing out I'm going to get from our Cluster director about the wreck." She shook her head. "Maybe I'm not cut out for this officer stuff."

"What's that supposed to mean?" said Kinet. "You're doing a great job."

"I don't know. I guess I'm just frustrated. This trip on the *Wandering Star* was supposed to be my first real command. It was also supposed to be routine: out to one of our regular trading partners, then straight back, an eight-month duty posting. That was before we got stranded here in Earth's solar system. We've been here almost a full year."

"Yeah," said Stas. "It's hard. We're learning so much, but I'd love to be able to go home. I wonder how the repairs are coming."

"I don't know," said Teract. "The big structural stuff is almost finished, but who knows about the rest."

"I wish I could even just *call* home," said Modulas. "My family is probably worried half to death about me."

"Whoa!" said Kinet, waving all four arms in a gesture of negation. "If we start talking about things like that, we're all just going to get depressed. I vote we change the subject."

"Have you got any bright ideas on what we should change it to?" asked Teract.

"Sure." Kinet grinned. "I think we should talk about my brilliance in coming up with that signal balloon."

"Sounds like someone's feeling full of himself," said Modulas.

"Or maybe just full of hot air," said Teract.

"Hey!" said Kinet. "None of that. It *was* brilliant. You have to admit that."

"Smart, yes," said Stas. "Brilliant, no. You were just applying simple principles that have developed over the centuries on most planets."

© It's About Time

"What?" said Kinet. "No way."

"Sure," replied Stas. "My people. Yours. Humans too. Most intelligent species have thought about the nature of matter and how heat and cold or other changes affect it."

"I don't believe it," said Kinet. "Not centuries. Take humans. They're much smarter now than their ancients were. Look at all the things they know that their ancestors didn't."

"That's getting knowledge confused with intelligence," said Stas. "Both the ancient Greeks and the Chinese thought about the nature of matter and tried to build mental models of how it all worked. Admittedly, they came up with ideas that we can now see were too simple, but that's not because they weren't smart. It's mostly because they didn't have anyone else's ideas to build on or the logical or physical tools to really explore their ideas. Even without a tradition of scientific exploration and without microscopes or other tools, they tried. They came up with a simplified list of elements from which they thought all matter was made. The Greeks had four basic elements, the Chinese five."

"That's awfully short of the hundred or so we currently know about," said Kinet.

"Sure," agreed Stas, "but everyone has to start somewhere."

She'd have said more, but the intercom clicked on. "Splintership Commander Teract Barenduin, report to your Cluster director immediately."

"Oh gralf!" said Teract. "That didn't take very long. Maybe I should have reported in when I first got here. Better that the director heard the bad news from me than that she got it second hand. I'm going to get roasted." Teract climbed to her feet and straightened her jacket. "Wish me luck."

✳ ✳ ✳

"Teract's been gone an awfully long time," said Stas. "I hope Cluster Command isn't chewing on her too much."

"Me too," agreed Kinet. "Sure she drove the boat into a log, but anyone could have done that. It's what you do *after* the disaster that shows real leadership, and she did great."

"Yes, she did," agreed Modulas. "She really knocked herself out making sure we all made it safely to shore and diving for our equipment afterward."

"She certainly doesn't deserve a reprimand and …" Stas's words were cut off by the sound of an opening door.

Teract had returned and her expression was very grim. The corners of her mouth, normally showing at least the hint of a smile, were turned sharply down. Her tail, a fine indicator of her moods, hung limp. Even having it lash back and forth in anger would have been better than that. For a long moment, no one said anything. Nobody wanted to be the one to ask how it had gone.

© It's About Time

Finally, Kinet got up the courage to ask, "What happened? Were they really hard on you about the wreck?"

"It didn't come up," said Teract. She dropped into a chair and slid out of her formal uniform jacket. The tank top underneath was still filthy from their recent stranding on Earth, but it felt far more comfortable.

"It didn't come up?" said Modulas. "Why are you looking so down then?"

"The Dreamers are shot."

"What?" asked Kinet.

"The Dreamers are the computers that control the *Wandering Star's* nanobots," interjected Stas.

"I know that," said Kinet. "It's the shot part that's got me worried. Are they really gone?"

"Fried," said Teract. "Cooked. Totally fragged."

"That's very bad," said Modulas. His skin had begun to glow brightly as the implications sank in. The sub-microscopic robots called nanobots were essential to replacing the Fold Space Lance which they needed to get home. The Dreamers were the master controls for the nanobots. No nanobots meant no Lance heads and the Lance was the device that made it possible to break into Fold Space and travel faster than light. Without them, it would take the *Wandering Star* a hundred thousand years to get home.

"What happened to the Dreamers?" asked Stas.

"The cryogenic cooling system ruptured," said Teract. "Remember when we made our crash breakthrough into Earth space and we had that big power loss?"

"Yes," said Kinet.

"Apparently, it hit the cooling systems especially hard. The cryogenics were without power for 15 hours. During that time, the cooling fluid overheated and turned back into a gas."

"That shouldn't have been a problem," said Modulas. "Sure the stuff expands, but that's what overpressure valves are for. That's why they put those little gaskets on pressure cookers."

"They failed," said Teract. "There was some sort of contaminant in the lines, tiny particles of some reactive metal that built up on the valve surfaces. Under normal circumstances, it's no problem, but when the system heated up so fast, some of the extra energy went into those particles. They bonded with the valves and chemically welded them shut."

"Huh?" said Kinet.

"I don't entirely understand it myself," replied Teract. "The tech who explained things told me to think of the model of a red and green candle sitting in a box that got left in the Sun. The candles are stable sitting on your desk, but when you add the heat energy from the Sun, they melt and flow together into a puddle. All of the original stuff from the candles is still there, but it's now become one thing. Where there wasn't much contact, there may still

© It's About Time

be bits of red and green wax that could be separated pretty easily. At the points where the two colors have really bonded, you get a mix that's a new color and it's very hard to get back to your starting point. The contaminant and the valve surfaces flowed together like the candles, and that kept them from opening."

"That sounds ugly," said Modulas.

"It was," said Teract. "Instead of blowing out the valves, there were multiple line ruptures when the stuff expanded. Pop. Pop. Pop." She cracked her tail like a whip to emphasize the pops. "They were all through the main nexus of the Dreamers and they tore the organic nanocircuitry to shreds."

"But you need hordes of nanobots and careful control to repair microcircuitry like the stuff in the Dreamers," said Kinet.

"Yes," said Teract, "and you need the Dreamers to run the nanobots. So unless someone thinks of something really clever, we may be staying here and studying the Earth for a *very* long time. The systems involved are just too small for anyone to work on without using the nanobots."

"You can say that again," said Modulas. "Some of the circuits in the lances are only one or two particles wide, and there are thousands of them."

Kinet shook his head. "How hard can it be, really? We do circuit repair all the time. I helped you rewire your suit when it was having problems after the *Solar Wind* fell on it. That wasn't so bad."

"Are you kidding?" asked Modulas. "Don't you know how small these thing are?"

"I'm a pilot. What do I know about small? Help me out here. I mean I've seen the lances. They're about six meters long, and as big around as my arm. I know everyone says that the built-in circuits are on an atomic scale of size, but now I'm not sure what that really means. I guess I always just figured that meant really small."

Modulas reached over with one of the robotic hands on his suit. He pinched the hair on Kinet's arm gave it a sharp yank.

"Hey!" exclaimed Kinet. "That hurt."

"I needed a hair," said Modulas.

"Well, you got a lot more than one," replied Kinet, pointing at the small clump of hairs between Modulas' mechanical fingers.

"Sorry about that, these suits aren't great on fine control. I should have had you do it. Too late now, I guess. Take one of the hairs." He extended his hand to Kinet and the four-armed alien picked one up.

"Got it," said Kinet.

"Look at it closely," said Modulas. "What do you see?"

Kinet squinted. "Not much. It's too small for me to make out any real details."

© It's About Time

Modulas nodded. "Exactly, and on the atomic scale of the lance circuits, that hair is huge."

"Really?

"Really," said Modulas. "Imagine that hair was as big as a freeway with four lanes of traffic moving in each direction."

"So, eight lanes total," said Kinet, doing a quick calculation in his head. "That's almost 35 meters." Then he held the hair up. "Think of this hair being as big as a freeway?"

"Yes," said Modulas. "Now imagine a miniature you and me driving a car on that freeway."

"All right." Kinet nodded.

"Say that the miniature me reached over to the miniature you, like this." He grabbed another bit of Kinet's hair.

"Don't!" squeaked the furry alien.

"Imagine that the miniature me grabbed a hair," said Modulas, releasing his grip. "Then, he handed the miniature you the miniature hair to look at. That teeny tiny hair would still be more than twice as big across as a single atom."

"An atom is smaller than a hair from a me little enough to drive a car on a normal hair?" asked Kinet, looking at the hair in his hand.

"Yes," interjected Teract, leaning in to join the conversation, "and the computers that run our nano-machines, the only devices we have that are small enough to work with things on that scale, are fried."

"We have a serious problem," said Kinet.

Teract nodded glumly in agreement.

Questions

1. Stas claims that Kinet is confusing knowledge with intelligence. In other words, Kinet apparently believes that people on Earth have only recently become intelligent. Do you agree with Stas or Kinet? Explain why or why not.

2. When the tech explained the problems from heating the system too fast to Teract, why did he mention candles?

3. Teract reports, "The cryogenic cooling system ruptured." What does cryogenic mean? Where could cryogenics be used in the real world?

© It's About Time

© It's About Time

Big Problems, Small Solutions

Teract sighed as she entered the lounge. "Well, that was useless." She had just come from a meeting of the Cluster Command. It had been called for discussion about the problem of the broken Dreamers, and was the 10th meeting in 10 days.

"Nothing new?" asked Kinet.

"Not a thing," she replied, her tail beginning to lash. "We just went back and forth over the same ground for hours. If it's left up to Cluster Command, we're never going to get home."

"I guess we can't leave it up to Cluster Command then," said Modulas.

"What?" asked Teract. She cocked her head to one side. "What do you mean? It's not like *we* can do anything." She toed a control on the nearest chair, changing its shape to fit her, tail and all, and sat down, eyeing her crew alertly.

"Actually," said Kinet, "Stas has a couple of ideas on that."

"All right," said Teract. "Let's hear 'em." She didn't really hold out much hope that Stas had found something that everyone else had missed. The best engineers in the fleet had been working on the problem almost since they arrived in Earth space. On the other hand, she figured that any suggestions for action would make a nice change from all the grumbling and shouting at the Cluster meetings.

"Well…" said Stas, tapping her long toes nervously on the carpet, "it's like this. When we haven't been studying the small humans, I've been digging around on the internet."

"Oh, no," said Teract, letting her head fall back against the chair's head rest. The last time Stas had suggested a solution from the internet, it had led them to Alphonse I, King of the Earth. Since the Earth really didn't have a king, the incident made Teract and her fellow crew members from the *Solar Wind* laughingstocks of the fleet.

"It's not like that this time," said Stas, understanding Teract's reaction. "I promise."

"She's right," said Kinet. "I don't know if what she's suggesting will work, but it's better than anything else I've heard."

"All right," said Teract, "go ahead."

"Right," replied Stas, her toes still madly tapping away. "I'll have to fill in a little background here. Earth's scientists have just begun to work with nano-sized machines. They're very basic, just a gear here or a lever there or at most, a couple of tiny motors, at least so far. But humans have written a lot of stuff called science fiction, some of which talks about all the things nanobots *could* do."

"Got it," said Teract. "They've got nothing that will actually work, but they have made up a bunch of stories about fantasy machines that might help us if they existed. What's the point?"

"Hang on," said Modulas. "Let her finish."

"Sorry," said Teract. "I don't really mean to be rude. It's just I've had a long hard day, and I've been leery of the internet ever since that King of the Earth mess."

© It's About Time

"Me too," replied Stas, sharply snapping her mandibles together once for emphasis. "That's why I was very careful this time. I checked these stories out really thoroughly, and they're basically all wishful thinking." Teract opened her mouth again, and Stas hurried on. "But, and this is a big but, they have spent an awful lot of time thinking about how the technology could be used if they had it."

"Yeah," said Kinet, "it's like you said last week. Just because they don't have as much knowledge as we do doesn't mean they aren't as smart. They've thought about all the stuff we do with the nanobots and more things besides, even if they can't do any of it yet."

"What?" asked Teract. "Like building a Fold Space Lance?"

"Not in quite that detail," said Kinet, "but the principles are the same. We all know how it works. With the right programming, nanobots can build just about anything by chemically assembling the particles of the hundred or so elements into more and more complex structures. It just has to be put together properly. Humans often use the analogy of building a house. Start with a few lumps of metal ore and  some trees, shape and refine the materials into tools and nails and boards and things, and pretty soon you've got a house that you can live in."

"I know all the basic analogies," said Teract. "The wood and metal are like the atoms that make up the elements. By putting atoms together like simple but tiny building blocks, you can make anything in the universe. You just have to know how the different particles will interact to form new things. But I don't see how any of that helps us use our *real* nanobots to build *working* Lances so we can go home."

"I didn't at first either," said Stas. "Then I started reading about the many different ways the humans in the stories controlled their nanobots. In a couple of stories, they had computers like our Dreamers. The computers would lay out tiny frameworks of particles and sub-microscopic structures to build their devices."

"How else would you do it?" asked Teract. "There are so many possible ways to combine all the different particles that a computer is necessary to figure it all out. If you put this little bit of stuff together with that little bit, does it explode or does it form something new? Does it take energy to make the connections? Or do you get energy out of the interaction? It's all so complex!"

"Sure," agreed Stas, "but in a lot of stories, they just used computers in the design phase to figure out how all the bits go together. For the actual construction, they had humans guiding the machines. They operated the nanobots as if they were little tiny spaceships using control consoles very much like the ones we use on the *Solar Wind*."

© It's About Time

"But that's crazy," said Teract. "Why use people on a job like that? It would be tedious and difficult and it would take practically forever. If you have something like the Dreamers, it's a huge waste of personnel hours."

"But what if you don't have Dreamers because the cryogenics are cooked?" asked Kinet.

"Well…" said Teract, beginning to like the idea.

"And," added Modulas, "what if you have a whole bunch of trained personnel just sitting around doing nothing because they can't go home?" He tapped the controls on his internal computer, quickly moving the arms of his suit through a series of complex gestures. "What if those personnel just happened to be experienced in remote machine operation?"

"I'm still not sure," said Teract. "The Lances are really complex microstructures. They've got all kinds of miniature carbon wires and weird alloys…"

"Portions of three Lances survived the accident," said Stas. "I checked the data base and it looks as if we have one almost complete template to work from between the three sets of fragments."

"If we really do have most of a template, we'd just be doing a really elaborate copy job," said Teract, half to herself. After a moment, she looked up at Stas, and nodded. "It just might work. Those writers you looked at must be pretty good scientists."

"Actually," she replied, "many of them aren't scientists at all, they're just regular folks with an interest in science."

"Really?" said Teract. Then she shook her head. "I guess I shouldn't be surprised, since I know that anybody can learn science. It's a good idea, Stas. I'll take it to Cluster Command." She raised a warning finger. "But nobody mentions the internet or science fiction outside this room, right? After what happened the last time, that's a sure way to get the plan killed."

The other three aliens solemnly agreed.

✳ ✳ ✳

One week later, the crew of the *Solar Wind* was sitting at their command posts aboard the little splintership. The suggestion had gone all the way up the chain of command to the *Wandering Star's* governing council. The council decided it was an operational question instead of a research or contact issue and sent it to the ship's captain. She decided the scheme was worth a try. So the *Solar Wind's* crew, as the originators of the idea, got to try a first test run.

Teract had decided, with the other three concurring, that since Modulas had the most experience with remote operation of equipment he would control the construction parts of the nanobot swarm on this first attempt. Kinet would guide the tiny machines from point to point. Stas, with her fine eye for detail and careful study habits, would pick out the work site and monitor their progress from a slight distance to make sure they kept on track. Teract, as always, would see to it that everyone worked as a team and she would make any command decisions that needed to be made.

"Operational status?" asked Teract.

© It's About Time

"Guidance controls ready," said Kinet.

"Assembly and fabrication units ready," said Modulas.

"Observation and verification systems engaged," said Stas.

"Let's do it," said Teract. "Open phase one."

Kinet hit a button and the image on Teract's screen began to change. Her computer was hooked into a powerful sensor system in the nanobot labs located in another part of the ship. The sizes involved would be too small for an optical microscope to see, so the image on Teract's screen was computer generated rather than an actual picture. But the systems involved were so good that no one could have told the difference.

For now, Teract's view looked down on a one meter segment of Fold Space Lance floating in a vacuum tank. Her apparent distance from the piece of broken equipment was about two meters. The Lance piece came from close to the tip. It looked a bit like a chunk of an Earth plane's propeller, a rounded blade shape that spiraled slightly. It appeared to be made of some kind of shiny black glass or stone with a dusting of gold sparkles.

That view held for a few seconds until Teract spoke again. "Phase two."

Instantly, her point of view swept in closer to the Lance's surface, increasing in magnification. The propeller-like blade seemed to grow, becoming an avenue of black glass laced with thousands of tiny threads of gold. The magnification increased again and the *Solar Wind* appeared to be hovering above a huge obsidian plain. Golden ropes seemed to be tracing intricate patterns just under the surface, some of them plunging deeply out of sight. A cloud of gnat-like shapes hovered above the surface. They were so tiny that even at this magnification, they could only be seen because there were so many of them.

Once more the magnification increased. Now, the golden ropes, which were each actually less than a 10th the size of a human hair, appeared large enough for a city bus to drive on. It became clear that they too were covered in intricate patterns of other materials, like vines of platinum and black silk almost too fine to see. The gnat-like cloud resolved into thousands of individual shapes, still little bigger than mosquitoes. The next magnification turned the fine tracery of vines on the gold strands into a giant spider's web. The mosquitoes became jeweled hawks—all angles and edges. These were the nanobots. Their job was to analyze the delicate traceries of materials inside the gold threads and to lay down a framework to duplicate those traceries mechanically and chemically.

Magnification increased one final time, and now the nanobots appeared to be huge construction machines with lifters and grabbers, blades for pushing and hooks for pulling. From Kinet's seat, he controlled the movement of all the thousands of little machines. He made sure that none of them collided and that each one was in the proper place for Modulas to guide its tools. Stas' view screen was split, showing both this view and a second vacuum tank where another swarm of nanobots hung waiting in empty space.

"Phase three," said Teract.

The nanobots began to slowly strip away the surface of the nearest golden thread, atom by atom. As they did so, Stas used the computer to note the position of each atom, so that the proper framework could be built in the other tank for the test run. After three hours,

© It's About Time

the aliens stopped their work. In the second tank, a skeletal framework of particles and support structures had taken shape. The next step was to bathe the framework in a series of carefully formulated chemical baths over several days. In the baths, the particles on the lance would chemically bond with other carefully selected particles to form new molecules. Some of the mixes would have to be heated or electrified to facilitate the right chemical interactions. Others would actually generate energy as a byproduct of their interactions, like a battery. With each immersion, another type of particle would be added to the mix until finally, if everything worked right, they would have recreated a cubic centimeter of the Lance. Only then would they know if the technique would work and if they'd ever be able to go home.

Questions

1. Science is more than just knowledge; it is also creatively using that knowledge by connecting it together in new ways. Which alien made the creative jump in this story, and what allowed the alien to do that?

2. Teract says that anyone can learn science. However, learning can mean learning about science or learning to do science. Which meaning do you think Teract meant when she used that phrase? Explain.

3. In your or your parents' lifetime, what are some devices that people used to think of as just "science fiction" that have actually been made in our world?

© It's About Time

© It's About Time

Science and Home

The Lance fragment that the crew of the *Solar Wind* had so carefully assembled was almost complete. It went through some final chemical processes, laser etching, and electroplating. It then went to the lab for a vigorous round of tests. Eight days later, the results came back. The lab's scientists thought the fragment looked and acted like the real thing. Whether a complete Lance head built in the same way would work as well as the original remained to be seen. Before they could be certain, the device would have to be rebuilt from base to tip and processed all at once. It was a much more complex task. Each Lance head would take months of work by hundreds of crew members. Still, the news called for a celebration, so the four aliens reserved a table at the Looking Glass.

The restaurant was one of the nicest on board the huge Trade Confederation vessel *Wandering Star*. It was located in the ring section of the ship and its floor was made of clear armorglass. Since the ship constantly rotated, this gave the diners a spectacular and constantly changing view of the Earth-Moon system and the local starscape. So the foursome put on their best dress uniforms for the occasion, and treated themselves to a very expensive dinner. After dinner, they sat around sipping their drinks and wondering about the next few months.

"I can't believe it worked," said Stas, shaking her head.

"Why not?" asked Kinet. "It was your idea, after all."

"Maybe that's it," she replied. "It's a lot easier to believe in a solution when it comes from some genius or an authority figure. When you think of the kind of person who solves important scientific or engineering problems, you think of someone famous and distinguished like Earth's Einstein or Marie Curie."

"But that's not really the way it works," said Teract. "Anyone can do science, and the more people who try it, the more that will be accomplished."

"I know," said Stas. "It's just hard to remember that sometimes."

"Well," said Kinet, "try not to forget again for the next couple of months. We have a lot of hard scientific work ahead of us if we want to get home."

"That's the truth," said Teract. "Let me propose a toast. Science and home!"

"Science and home!" agreed the others as the four clinked their glasses together and drank.

✳✳✳

As the *Wandering Star* prepared for Fold Space breakthrough, the tension aboard the Trade Consortium ship grew. Nowhere was that more true than in the cabin of the *Solar Wind*. The idea for how to replace the ruined Fold Space Lance heads had come from the four young aliens who made up the splintership's crew. If anything went wrong, the ultimate fault would be theirs, and it would be spectacular. The replacement heads for the Lance had

© It's About Time

checked out all right, but only individually and only in practice. The true test would come when they actually attempted a Fold Space crossover, an attempt that could lead to utter disaster. If the Lance failed, the best that could happen would be that the system simply didn't fire. If so, there would be no hole in the fabric of space-time and the *Wandering Star* and all of its people would simply remain trapped in near-Earth space.

That was only one possible outcome, and the least likely one at that. Tests had proven that the Lance could be fired, and that meant that it would almost certainly punch a hole in the universe. If that hole wasn't big enough, disaster would follow. Moving at 90 percent of the speed of light, the ship couldn't possibly turn aside in time. Instead, the ship, or at least parts of it, would go right on through. Anything that didn't fit the size and shape of the hole, however, would be sliced off and left behind. At that point, it would be a tough call. Who was worse off: the aliens who went with the main part of the ship or those trapped in the wreckage left behind?

The Lance heads were mounted on the outermost edges of the ship. They would certainly be stripped away in any accident, and those who made it through would be stuck in Fold Space forever. Those left behind would be moving away from Earth at nine-tenths the speed of light. Without the massive main engines of the *Wandering Star* to use as brakes, it could take them 20 years to come to a stop. By that time, they'd be a long way from the planet that had been their temporary home. Even if the breakthrough itself went perfectly, the Lance heads might burn themselves out in the process, trapping the whole ship in Fold Space with no way to get back.

Every possible variation of disaster played itself out in the minds of the four friends as the ship began its acceleration for departure. Too tense to talk, the four faced the future each in his or her own way. Modulas ran endless engineering simulations on his suit's computer. His expression remained stoic, but his skin glowed so brightly from worry that none of the others could look at him for long. Stas sat in her backward facing chair, eyes fixed on the systems monitor. Her chiton features couldn't show much emotion, but the rasp of her toes clenching and unclenching spoke for her. Kinet fidgeted, keeping all six limbs in constant motion. Even his ears and nose twitched occasionally. Only Teract appeared fully calm, her face relaxed, her hands loose on the chair's arms, her legs casually crossed in front of her. Anyone sitting behind her though would have seen that the only way she'd kept her tail from lashing was by wrapping it three times around the leg of her command chair.

When the ship reached 60 percent of light speed, the Fold Space Lance began its charge cycle. At 70 percent, it finished cycling and the four Lance heads went active. Almost immediately, the control light for the number three Lance head blinked over from the green of full readiness to the blue that signaled a slight irregularity in its status. The four held their breath, wondering if the captain would order an abort. They knew that under normal circumstances, any three heads would be plenty and such a small flutter wouldn't justify a shutdown, but these were anything but normal circumstances. The abort didn't come, and the *Wandering Star* continued to pick up speed.

© It's About Time

At 80 percent of light speed, the order to engage the Lance came down the chain of command. With a noise like tearing sheets amplified a thousand times, the Lance heads fired. From each of the four heads a burst of energy roared toward an invisible point in space. Three of the beams were crisp and straight, exhibiting the pure violet of proper function. The fourth, coming from the number three head, was ragged with an ugly blue undertone that colored the edges of the energy stream. At an invisible point in space the four beams met and the stuff of the universe began to bulge and tear. The *Wandering Star* hit point eight-five light-speed, point eight-six. Eight-seven. They were nearly there. A sharp frightening noise came through the very structure of the ship, more felt than heard, and the number three Lance head cut out entirely.

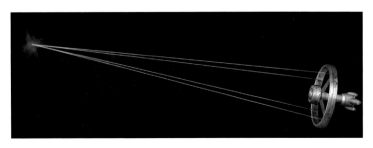

Still, where the remaining blasts met, the fabric of space and time had already endured unimaginable stresses. Finally, the Lance heads tore an opening into Fold Space. The gap only held for a few seconds, but the *Wandering Star* was approaching it at 90 percent of the speed of light. The four aliens watched the distance narrow with a growing sense of dread. The hole didn't look as big as it should have, but it was too late to turn aside and the ship plunged ahead. In an instant, it was all over. The huge ship hit the hole and vanished from the normal universe. Behind, a single sliver of hull metal forty meters long, one meter wide and a single millimeter in thickness continued along the ship's original course, shaved off by the edge of the hole. It had been that close, but they were through.

The time to relax had not come yet though. All motion and sound ceased in the cabin of the *Solar Wind* as they waited to hear the system status reports. The *Wandering Star* had barely cleared the edges of the crossover point, and they wouldn't have made it at all if the number three Lance hadn't at least partially fired. Clearly the nanobot control system they'd devised had not met with full success. If the malfunction that had caused number three to shut down turned out to be permanent, they might not be able to make it back out of Fold Space at the other end of their journey.

"Lance head one, function stable at 89 percent power," read Stas as it came across the system status screen.

"Lance head two, function erratic, variance between 92 and 95 percent power."

"Lance head four …"

Kinet cut her off. "What about three?" he asked, tapping at his own computer.

The answer flashed across the screen and he repeated it aloud, "Data not available, diagnostics still cycling."

"Gralf," muttered Teract, her voice barely audible. "Well, let's hear from four."

"Lance head four, 100 percent function."

"That's something at least," said Modulas. "I wonder how long before we hear about number three."

© It's About Time

The command deck speaker gave the standard signal for a shipwide address from the captain. All four aliens turned to look at it as the captain's clear soprano rang out, "All right, people, I've just gotten the final word from the techs on Lance head three. The power loss was due to a faulty relay coupling. They tell me that they'll have it back on line in about three hours and that the head should be functioning at a stable 76 percent of capacity. It looks like we're going to make it home after all."

On the deck of the *Solar Wind*, the silence that followed lasted for about 10 seconds before it vanished in wild cheers and whoops, and tears and laughter.

Questions

1. Stas says, "It's a lot easier to believe in a solution that comes from some brilliant authority figure." Teract replies, "But that's not the way it really works…". Where do you think solutions come from in science? Explain.

2. Teract states that she believes that "anyone can do science." What qualities would such a person have or need to develop to "do science?"

3. What are your ideas about how the *Wandering Star* could be better designed to ensure that the ship would still make the journey home, even if something went wrong?

Even aliens need some down time! Stas, Teract, Modulas and Kinet also known as "The Alien 4s" enjoy getting together for a jam session every chance they get.

© It's About Time